Estimating Health Risks From Infrastructure Failures

Awwa Research Foundation

About the Awwa Research Foundation

The Awwa Research Foundation (AwwaRF) is a member-supported, international, nonprofit organization that sponsors research to enable water utilities, public health agencies, and other professionals to provide safe and affordable drinking water to consumers.

The Foundation's mission is to advance the science of water to improve the quality of life. To achieve this mission, the Foundation sponsors studies on all aspects of drinking water, including supply and resources, treatment, monitoring and analysis, distribution, management, and health effects. Funding for research is provided primarily by subscription payments from approximately 1,000 utilities, consulting firms, and manufacturers in North America and abroad. Additional funding comes from collaborative partnerships with other national and international organizations, allowing for resources to be leveraged, expertise to be shared, and broad-based knowledge to be developed and disseminated. Government funding serves as a third source of research dollars.

From its headquarters in Denver, Colorado, the Foundation's staff directs and supports the efforts of more than 800 volunteers who serve on the board of trustees and various committees. These volunteers represent many facets of the water industry, and contribute their expertise to select and monitor research studies that benefit the entire drinking water community.

The results of research are disseminated through a number of channels, including reports, the Web site, conferences, and periodicals.

For subscribers, the Foundation serves as a cooperative program in which water suppliers unite to pool their resources. By applying Foundation research findings, these water suppliers can save substantial costs and stay on the leading edge of drinking water science and technology. Since its inception, AwwaRF has supplied the water community with more than $300 million in applied research.

More information about the Foundation and how to become a subscriber is available on the Web at **www.awwarf.org**.

Estimating Health Risks From Infrastructure Failures

Prepared by:
Karen M.E. Emde
Capital Health, Edmonton, AB, T5N 3G1

Daniel W. Smith
Department of Civil and Environmental Engineering, University of Alberta, Edmonton, AB, T6G 2G7

James A. Talbot
James A. Talbot Professional Corporation, Edmonton, AB, T6E 2L3

Les Gammie and **Susan Ancel**
EPCOR Water Services Inc., Edmonton, AB, T5J 3B1

Nelson Fok
Capital Health, Edmonton, AB, T5N 3G1

and

Janet Mainiero
New York, NY 10520

Sponsored by:
Awwa Research Foundation
6666 West Quincy Avenue, Denver, CO 80235-3098

Published by:

Awwa Research Foundation

American Water Works Association

IWA Publishing

DISCLAIMER

This study was funded by the Awwa Research Foundation (AwwaRF). AwwaRF assumes no responsibility for the content of the research study reported in this publication or for the opinions or statements of fact expressed in the report. The mention of trade names for commercial products does not represent or imply the approval or endorsement of AwwaRF. This report is presented solely for informational purposes.

CONTENTS

TABLES

FIGURES

FOREWORD

The Awwa Research Foundation (AwwaRF) is a nonprofit corporation that is dedicated to the implementation of a research effort to help utilities respond to regulatory requirements and traditional high-priority concerns of the industry. The research agenda is developed through a process of consultation with subscribers and drinking water professionals. Under the umbrella of a Strategic Research Plan, the Research Advisory Council prioritizes the suggested projects based upon current and future needs, applicability, and past work; the recommendations are forwarded to the Board of Trustees for final selection. The foundation also sponsors research projects through an unsolicited proposal process; the Collaborative Research, Research Applications, and Tailored Collaboration programs; and various joint research efforts with organizations such as the U.S. Environmental Protection Agency, the U.S. Bureau of Reclamation, and the Association of California Water Agencies.

This publication is a result of one of these sponsored studies, and it is hoped that its findings will be applied in communities throughout the world. The following report serves not only as a means of communicating the results of the water industry's centralized research program but also as a tool to enlist the further support of the nonmember utilities and individuals.

Projects are managed closely from their inception to the final report by the foundation's staff and large cadre of volunteers who willingly contribute their time and expertise. The foundation serves a planning and management function and awards contracts to other institutions such as water utilities, universities, and engineering firms. The funding for this research effort comes primarily from the Subscription Program, through which water utilities subscribe to the research program and make an annual payment proportionate to the volume of water they deliver and consultants and manufacturers subscribe based on their annual billings. The program offers a cost-effective and fair method for funding research in the public interest.

A broad spectrum of water supply issues is addressed by the foundation's research agenda: resources, treatment and operations, distribution and storage, water quality and analysis, toxicology, economics, and management. The ultimate purpose of the coordinated effort is to assist water suppliers to provide the highest possible quality of water economically and reliably. The true benefits are realized when the results are implemented at the utility level. The foundation's trustees are pleased to offer this publication as a contribution toward that end.

Walter J. Bishop
Chair, Board of Trustees
Awwa Research Foundation

Robert C. Renner, P.E.
Executive Director
Awwa Research Foundation

ACKNOWLEDGMENTS

The workshop and resulting manual were developed with the help of numerous individuals and agencies that openly shared their experiences, thoughts, and expertise, resulting in the development of a document beneficial to public health agencies and water utilities. This exchange helped create international partnerships between complementary agencies and friendships between individuals, resulting in an enhanced understanding of international and transcontinental health and drinking water issues.

The authors of this report are indebted to the following water utilities, public health agencies, police department, private consultants, and academic researchers for their open, enthusiastic cooperation, guidance, wisdom, and participation with the various requirements and tasks required of this project:

- Agency for Toxic Substances and Disease Registry, Atlanta, Ga., United States, Morris Maslia
- AH Environmental Consultants Inc., Hampton, Va., United States, David Hiltebrand
- Alberta Environment, Edmonton, Alta., Canada, Pat Lang
- Alberta Health and Wellness, Edmonton, Alta., Canada, Kevin McLeod
- Alberta Provincial Laboratory of Public Health, Edmonton, Alta., Canada, Edie Ashton
- Boston Public Health Commission, Boston, Mass., United States, M. Anita Barry
- Boston Water and Sewer Commission, Boston, Mass., United States, William Foley and John Sullivan
- Capital Health, Edmonton, Alta., Canada, Gerry Predy
- Centers for Disease Control and Prevention, Atlanta, Ga., United States, Deborah Levy
- Cincinnati Department of Water Works, Cincinnati, Ohio, United States, David Hartman
- City of Edmonton Police Services, Edmonton, Alta., Canada, Mike Cook
- City of New York Department of Environmental Protection, New York, N.Y., United States, David Lipsky
- Cooperative Centre for Water Quality and Treatment, Salisbury, South Australia, Australia, Don Bursill
- E. Geldreich Consulting Microbiologist, Cincinnati, Ohio, United States, Edwin Geldreich
- EPCOR Water Services Inc., Edmonton, Alta., Canada, Susan Ancel and Gabrielle Deagle
- Health Canada, Edmonton, Alta., Canada, Brenda Woo
- Health Canada, Guelph, Ont., Canada, Jeff Aramini
- Health Canada, Ottawa, Ont., Canada, Paul Sockett
- Monash University Medical School and Cooperative Centre for Water Quality & Treatment, Melbourne, Australia, Martha Sinclair
- Seattle Public Utilities, Seattle, Wash., United States, David Muto
- Sydney Water, Liverpool, New South Wales, Australia, Philip Broad
- United States Environmental Protection Agency, Washington, D.C., United States, Kenneth Rotert

- United Utilities plc., Great Sankey, Warrington, Great Britain, Keith Osborn
- University of Alberta, Department of Civil and Environmental Engineering, Edmonton, Alta., Canada, Kyla Dahlgren and Clark Svrcek
- University of Alberta, Department of Public Health Sciences, Edmonton, Alta., Canada, Steve Hrudey
- University of Michigan, School of Public Health, Ann Arbor, Mich., United States, Rolf Deininger
- W.M. Grayman Consulting Engineer, Cincinnati, Ohio, United States, Walter Grayman
- Western Australia Water Corporation, Leederville, Western Australia, Australia, Richard Walker

Finally, the authors are deeply indebted to the enthusiastic, expert, and patient guidance of the Awwa Research Foundation project manager, Ryan Ulrich, and the Project Advisory Committee, consisting of Peter Chapman, Yakir Hasit, and Wayne Miller.

EXECUTIVE SUMMARY

The task of estimating health risks from infrastructure failures consisted of two components: a workshop, the purpose of which was to have water utilities and the various public health, environment, medicine, engineering, regulatory, and first-response agencies mutually explore the needs and capabilities of detecting, preventing, and correcting adverse health outcomes caused by infrastructure failures; and a writing effort to consolidate the considerable knowledge into this report. The results of the workshop that was held in July 2002 have been woven throughout every chapter of this document. The initial chapter presents the most salient points of the water system infrastructure, from its early development through the modern need and requirement for the establishment of working relationships and communication between utilities and public health agencies. Examination of this relationship yielded a matrix of health promotion, and primary and secondary prevention, verses basic objectives, the public's health, and environmental engineering.

The roles and responsibilities of water utilities were examined based on infrastructure failure conditions, vulnerability of distribution infrastructure, prevention of infrastructure failures, water quality monitoring, management of failures, and information systems or tools. The distribution system features that can impact public health were the design criteria, construction practices, locations of water mains within the system, ages of the components, and accessibility for operation and maintenance (O&M). Other issues were related to water utility O&M with respect to the frequency and types of efforts, water quality monitoring, and the use of supervisory control and data acquisition support. The people issues include education programs, attitudes, and communications. Several factors are beyond the control of water utilities; these include natural disasters, human-caused disasters, and public fears and rumors. Reduction of infrastructure failures after installation can be related to utility O&M efforts that include, but are not limited to, mains cleaning, hydrant and valve inspections, leak detection, transmission main inspections, reservoir management, cross-connection control programs, and monitoring. Management of infrastructure failure requires recognition of the types of failures—that is, structural, power, contaminant, and so forth, as well as the development of appropriate emergency plans for each type of event.

Public health agencies have a perception of risk based on the overarching responsibility to protect the health of the entire community. The use of risk assessment, in addition to risk analysis by the public health community, is based on the precautionary principle that "when an activity raises threats of harm to human health or the environment, precautionary measures should be taken even if all cause-and-effect relationships are not fully established scientifically" (Science and Environmental Health Network et al. 1998).

Active communication and collaboration are the methods that enable the entire process to work effectively. The framework and guidelines for pulling together a diverse and flexible team that can function during routine and emergency events are presented in chapter 4. Members for the water utility, public health agency, first responders, and elected officials are identified. Ideas for how to develop the team are presented, as well as how to improve communications through the effort of developing a common language and knowledge base. The types of problems associated with failure to manage risk were used to show the opportunities for application of these tools. The *Salmonella typhimurium* outbreak event in Gideon, Mo., and the chronic chemical contamination in Woburn, Mass., were studied for this evaluation. Both case studies led to attempts to learn from other industries using the concept of total quality management. This requires independent, as well as joint, surveillance systems that are capable of quickly identifying, monitoring, and correcting infrastructure failure events; introduction of waterborne contaminants and their zones of influence and exposures; identifying acute and/or chronic health consequences caused by failure events, together with monitoring and evaluation of these efforts to ensure successful interventions.

Communication, coordination, and collaboration among all of these groups is essential should there ever be a danger to public health as a result of an infrastructure failure. Developing and furthering these relationships, and building trust long before the failure event, ensures efficient and effective utility and public health intervention. However, the real goal is to build relationships while improving current practices and technologies. Cooperative and collaborative extramural efforts are vital for developing better, or new, detection technologies and for continuing to protect the health of the public and the community.

CHAPTER 1
INTRODUCTION TO HEALTH RISKS FROM INFRASTRUCTURE FAILURES

Karen M.E. Emde and Daniel W. Smith

HISTORICAL BACKGROUND

Access to safe, reliable, and adequate potable water supplies is the cornerstone for community development, growth, and public health protection (Antoun et al. 1995; Carmichael 1993a, b; Mays 2000). The establishment and growth of communities has historically been linked with access to adequate potable water, as well as the ability for remote disposal of community wastes (Carmichael 1993a, b; Prescott and Winslow 1904). As urbanization increased, so did the need to transmit water farther from a central supply to reach the entire community. The requirement to successfully meet this ever-increasing demand spurred numerous innovations in technology and practice in water treatment and water distribution, as well as in medicine and public health (McMichael 2000).

Early distribution systems included irrigation canals, cisterns, aqueducts, reservoirs, clay pipes, inverted siphons, and pressure pipes. The earliest distribution systems were constructed and used by the Greeks from approximately 6000 BC to 28 BC, then by the Turks from approximately 2000 BC to 200 BC, and also the Romans from approximately 312 BC to 226 AD (Mays 2000, Moyer 2003). The ancient distribution pipes of the Greeks and Turks were constructed mainly of wood or terra-cotta; however, the plumbing engineers of the Roman Empire refined the use of lead in pipes, solders, and fittings, and crafted the design and engineering technology that allowed for the construction of miles of aqueducts, delivering water to Roman cities from distant source waters (Mays 2000, Moyer 2003). During the Dark Ages, water infrastructure, as well as other early infrastructure, fell into disrepair and disuse. Several centuries passed and countless individuals died in epidemics from waterborne disease and other types of illness before sanitation and potable water again became a priority. Ancient engineering design and construction principles remained basically unchanged until the late 18th and early 19th centuries with the development of steam-driven water pumps and cast-iron water mains.

From the late 19th century to the present, huge strides have been made in technology and scientific knowledge, including engineering, science, medicine, and public health disciplines, such as the discovery that microorganisms are the cause of many infectious diseases. Scientific studies explored interactions among humans, their environment (e.g., chemical, biological, and physical factors), and microorganisms, hoping to gain a better understanding of the world around them. This spurred further improvements in sanitation and hygiene, water treatment and distribution practices, and control of many infectious diseases, as well as the development of new immunization and antibiotic therapy programs.

Enhanced public health surveillance, together with applied research and disease outbreak investigation, helped reveal sources of waterborne contaminants and modes of transmission; this led to improved and/or innovative methods for prevention, detection, and treatment of disease, including waterborne disease. Prevention strategies included innovation in technology for the design, construction, operation, and maintenance of municipal water treatment and distribution systems. During the 20th century, new health and environment regulations were introduced and are continually being updated to ensure the safe manufacture and delivery of potable water, as well as the protection of the public's health.

The ability to maintain an adequate quality and quantity of water, as well as the structural integrity of existing water infrastructure, often fell victim to recurring episodes of social, political, or economic upheaval and/or instability (Babbit and Doland 1949). Numerous disease outbreaks, either as a direct result of infrastructure breakdown and interrupted water supply or from increased pollution attributable to interruptions in community waste disposal, quickly followed episodes of water supply disruption (Babbit and Doland 1949; Carmichael 1993 a,b; Mays 2000). This continues to be a serious issue for nations today where natural disasters or conflict continue to occur.

DISTRIBUTION INFRASTRUCTURE AND SAFE DRINKING WATER

Most consumers in much of the developed world implicitly assume that water flowing from taps in their homes, places of business, or places of leisure is absolutely safe and risk-free. This often contrasts with the beliefs of individuals in the developing world, or developing areas of industrialized nations, where the quantity and quality of delivered water may be neither consistent nor meet regulatory standards.

Municipal distribution systems and associated private building plumbing systems are ubiquitous throughout any community and provide every person, industry, and institution connected to the municipal supply with potable water. Because the distribution system is not a "one-way" system, it can be vulnerable to posttreatment contamination within each municipality served. This may be due to accidental or deliberate disruption of the structural integrity of the distribution infrastructure, actual failure of one or more components, or accidental or deliberate contaminant intrusion from cross-connections and backsiphonage. Rehabilitation of existing infrastructure, as well as construction of new infrastructure, has been defined as the most significant expenditure for future water utility capital allocation (USEPA 1997, Davies et al. 1997).

Distribution Infrastructure Failure

In this report, *distribution infrastructure* is defined as the actual distribution pipes, including but not limited to steel, galvanized, cast and ductile iron, polyvinyl chloride, reinforced concrete, asbestos–cement, wood, and others, as well as pipe linings, pipe couplings, valves (i.e., gate, butterfly, wedge, pressure-reducing, backflow prevention, and others), pumps, power transmission units, controls, meters, storage tanks and reservoirs, municipal service lines, and hydrants. The water treatment plant also contains a significant amount of infrastructure, including tanks, filters, pipes, couplings, valves, pumps, meters, storage tanks, valves, and so forth. Failure of water treatment infrastructure, and/or treatment processes, typically results in systemwide problems; this report will not explore failure of those processes. There are many excellent Awwa Research Foundation (AwwaRF) and American Water Works Association (AWWA) publications that provide instruction and guidance for proper water plant operations and performance process monitoring (e.g., Kirmeyer et al. 2000, Letterman 1999, many of the AWWA manuals, etc). Chapter 2 provides an in-depth discussion of the causes of infrastructure failure and the management strategies employed to control and prevent future failures.

RELATIONSHIPS BETWEEN PUBLIC HEALTH AND INFRASTRUCTURE FAILURES

The terms *health*, *healthy*, and *public health* invoke a variety of responses and a plethora of definitions among individuals, as well as interpretations of those definitions. The word *healthy* is commonly interpreted as an absence of disease. This implies a continuum of states of being healthy, ranging from an apparent absence of disease or being physically fit, to some state of unwellness that spans the range from somewhat healthy to somewhat unhealthy, and finally to the ultimate state of unhealthiness, death (Basch 1990, Last 1987, Wilson and Cleary 1995, Pruess et al. 2002). Unfortunately, definitions, interpretations, and applications of the word *healthy* are infinitely more complex than the simple linear range just described.

In 1948 the World Health Organization (WHO) defined health as "...a state of complete physical, mental and social well-being and not merely the absence of disease or infirmity" (WHO 2003). Thus, public health consequences from infrastructure failures—such as fear and panic (i.e., damage to emotional health); loss of business or income (i.e., damage to emotional and social health); loss of fire protection (i.e., damage to physical health); loss of water for sanitation (i.e., damage to physical and social health); and others—must also be considered, not merely the absence of community waterborne or other types of disease. These subtleties become significant when trying to determine and evaluate the extent of exposure(s), the dose(s) of contaminant(s) received, and the subsequent health consequences when water distribution infrastructure has failed or been compromised.

At the most basic level, public health goals and objectives represent a collective community concern and response to avoid disease agents, disease-creating conditions, or any other health threats (Last 1987, Birch, Stoddart, and Beland 1998, Eisenberg et al. 2002, Fay 2001). The way in which public health goals are defined and refined over time is a function of all of the following elements:

- Community perception(s) of immediate to long-term dangers to their individual and/or collective health
- Local government structure, values, policies, agendas, and available resources
- Existing medical knowledge, resources, and detection and treatment technologies
- Community social and cultural values

Typically, disease and illness are not randomly distributed throughout a population but rather are the result of interactions among

- The natural environment, including characteristics of the community's geographic location such as air/soil/water quality, climate, altitude, vegetation, etc.
- Human characteristics and behaviors, including genetic adaptations and predisposition to disease or effects of disease-causing agents; cultural/religious beliefs, values, and traditions; individual and community socioeconomic characteristics and demographic mobility, etc.
- Community characteristics, including patterns of urbanization, industrialization, and community development; anthropogenic changes to the community's environment as well as to local and regional groups and institutions. (Ross and Mirowsky 2001, Last 1987).

Often there may be differing or contradictory perceptions about the significance of assorted public health and environmental issues, the interventions needed, and the changes required to governing policy, procedure, and regulatory reform. This is primarily due to differing cultural perspectives and values of individuals and key stakeholders living in the community (Arquette et al. 2002, Williams and Florez 2002, Evans and Stoddart 1990, Calderon et al. 1993). Thus, functional definitions used to describe disease or illnesses are, and will continue to be, a function of

- The time, place, culture, and descriptive language in which the disease is described
- The knowledge available to society and the importance placed on the disease syndrome relevant to the significance or impact on the community

The definition of disease or illness is not static. The societal and community importance allocated to diverse disease syndromes changes in response to two scenarios. In the first case, during the period of reduced incidence of a particular disease, which is often due to the efficacy of active public health control and/or surveillance programs, there may be a tendency toward diminished importance placed on mechanisms of control of that disease syndrome, as well as reduced ability and resources available to maintain all control measures needed. In the second case, when there are increased cases due to recurrence of a previously known disease, or new outbreaks caused by unknown or emerging pathogens, fear, panic, and rumor drive the public's reactions and, consequently, an immediate public health response is needed. The urgency of this need fuels "disease detectives" and regulators to search for the causative agent to quickly find an adequate treatment or cure. This dynamic relationship fuels improved disease control programs, enhanced efforts at disease detection and treatment programs, and innovation in the treatment and delivery of safe drinking water.

Multidisciplinary Nature of Public Health Protection and Safe Drinking Water

Public health improvement and protection is multidisciplinary; this effort involves many professions and disciplines that include, but are not limited to, public health and medical professionals, engineering professionals, scientists (e.g., biology, chemistry, microbiology, physics, and others), analytical laboratory staff, health and environment regulators, first-response agencies, and core community stakeholders.

These groups act in concert but are governed by independent training programs, have different professional requirements and expectations, and often work under different legislative acts and regulations that define regulatory expectations and jurisdictions. All have the common goal of protecting the community's health. The paths followed to achieve those objectives reflect the particular agency's culture and tradition, available expertise, jurisdiction, specific community requirements or needs as well as community expectations, available technology and funds, regulatory requirements, and, most important, an interagency collaborative vision and common effort.

Examples of human-made failures caused by adversarial, confrontational, or dysfunctional stakeholder relationships include but are not limited to

- Unequal access to health care, safe drinking water, and/or other health resources
- Treatment efforts that receive higher priority than prevention programs
- Lack of critical information or incorrect, incomplete, or misleading information

Table 1.1
Public health and environmental engineering approaches to waterborne disease protection

	Basic objectives	Public health examples	Environmental engineering examples
Health promotion	To have the individual assume responsibility for their own health, usually in partnership with appropriate stakeholder agency	Promotion of healthy lifestyles and living environments, etc.	Promotion of safe water handling practices (i.e., water storage, point-of-use devices, plumbing codes, etc.)
Primary prevention	To reduce the incidence of disease and/or premature death by reducing or removing causes or sources of disease	Immunization programs, outbreak surveillance programs, active treatment of disease, etc.	Control of water and wastewater treatment processes, control of water delivery processes and infrastructure, etc.
Secondary prevention	To detect disease early, before it has time to progress	Notifiable disease surveillance programs, sentinel disease surveillance programs, improved detection and screening methods, etc.	Water quality monitoring programs, infrastructure integrity monitoring programs, etc. Routine infrastructure operation, maintenance and renewal programs, engineered changes to water treatment and/or infrastructure design and operation requirements, etc.

- Inadequacies and/or deficiencies in core public health protection administrative programs
- Inappropriate or lack of staff training programs
- Breakdown in communications
- Inappropriate investment in technology that is too sophisticated to be regularly used or maintained by the agency or the community
- Duplication of expensive and sophisticated technology by multiple agencies in the same community
- Inappropriate or misguided agency and community priorities (McElroy and Townsend 1989, Last 1987, Basch 1990)

The four cornerstones of public health protection are health promotion, health education, primary and secondary prevention, and tertiary care. Table 1.1 contrasts the approaches taken by public health and environmental engineering in response to health promotion as well as primary and secondary prevention. These responses are not overlapping but are complementary in protecting the community from waterborne disease.

From the fundamental perspective of public health risk protection efforts, the degree of risk to human health from a known or suspected health hazard does not need to be at a statistically significant level to require political and/or community action (Last 1987). Adverse effect does not need to be proven absolutely; it merely needs to be suspected to cause a public health agency to become concerned and, consequently, act according to precautionary principles required for community health protection (Last 1987, 1998).

All drinking water treatment and delivery technologies must produce a product that meets the current regulatory standards for potable water and must be safe; safe being defined as a legislated, acceptable risk level for the majority of the population. These standards change as knowledge and technology change. Additional public health protection can be provided through increased preventive operating and monitoring processes at the water treatment plant and within the distribution infrastructure, but this approach is not without added cost—to the utility, to the consumer, and finally, to the regulators. Regulatory standards for potable water are designed to protect the health of typically healthy individuals; individuals with unusual or special health issues may need to take additional personal precautions.

Agency Involvement With Safe Water

Many levels of government (local, provincial/state, federal) and government departments such as environment, health, and others work together to ensure that drinking water is safe for the community and that the water treatment plant, municipal distribution system, and private building plumbing systems function properly. Functional authority to implement and enforce legislation and regulations for safe drinking water and public health protection is the responsibility of individual government departments or jurisdictions. Chapters 2 and 3 provide further discussion and information about the roles and responsibilities of the different stakeholder groups that are involved with safe drinking water and public health. Chapter 4 provides guidance on forming multidisciplinary community safe drinking water teams. Chapter 5 provides insights into how these teams can function on a practical level, in contrast with the development of community emergencies when the team members are not communicative or cooperative. Finally, chapter 6 discusses future directions and needs, drawing parallels from other industries to illustrate key concepts.

CHAPTER 2
ROLES AND RESPONSIBILITIES OF WATER UTILITIES

Les Gammie and Susan Ancel

INTRODUCTION

An important part of the responsibility of water utilities in reliable delivery of safe drinking water is proper management of the water distribution infrastructure. A properly managed system will have programs in place to produce safe drinking water and to manage and minimize infrastructure failure. The utility should also have documented plans for response to minor localized events that have relatively low risk as well as plans to manage major catastrophic events that have severe systemwide implications.

The focus of this chapter is discussion of the role of water utilities in preventing and managing water infrastructure failures. There is special emphasis on infrastructure failure events having public health impacts. This chapter does not present an all-encompassing discussion of the design and management of a water distribution system; excellent references are available from AwwaRF and AWWA. Sources include, but are not limited to, Chadderton, Christensen, and Henry-Unrath 1992; Kirmeyer, Richards, and Smith 1994; Deb, Hasit, and Grablutz 1995; Deb et al. 1998; Grayman et al. 2000; Kirmeyer et al. 1999; Kirmeyer et al. 2000; Cromwell et al. 2001; Pearson et al. 2001; Grigg 2002; and Snyder et al. 2002, which provide more comprehensive discussion of programs discussed in this chapter.

Six major themes are explored in this chapter:

1. Infrastructure failure conditions—common types of infrastructure failures that could be experienced by a water utility
2. Distribution infrastructure vulnerability—key characteristics of a distribution system that make it vulnerable to failures that may have community health consequences
3. Prevention of infrastructure failures—different strategies that could be implemented to prevent and/or reduce infrastructure failures
4. Water quality monitoring—critical components of a successful water quality monitoring program
5. Management of failures—ways to mitigate the impacts of particular failure events
6. Systems/tools—information system tools that utilities can use to manage their infrastructure

INFRASTRUCTURE FAILURE CONDITIONS

Infrastructure failures can be defined in many ways. For the purposes of this report, an infrastructure failure is defined as an infrastructure event that results in degradation of the quality and/or quantity of water delivered to the consumer. Failure events may be caused by individual and/or cumulative structural failures of infrastructure components or by hydraulic upsets that alter the flow regime within the distribution system. These failures or hydraulic upsets can be due to normal aging of infrastructure or can be caused by a catastrophic event such as a tornado, flood, or earthquake. Infrastructure failure may also be caused by human error during normal operation of the distribution system or from the willful intent to cause damage by means of vandalism, terrorism, or other types of motivated attacks.

Water infrastructure can experience many failure modalities that include but are not limited to

- Structural failures, including main breaks, pump failures, and reservoir failures. Typical causes are internal and/or external corrosion, hydraulic pressure transients, or power failures.
- Cross-connections that may occur through the improper installation of pressurized equipment within customer properties, improper installation of water mains, and low water pressures. High groundwater and proximity to sewer mains may become a cross-connection concern. Petroleum contamination of soils surrounding a water main is a serious concern, especially at rubber-gasket joints, which are especially vulnerable to hydrocarbon penetration. Reservoirs and standpipes are particularly vulnerable at the access hatches and vents.
- Intramural hazards, such as biofilm separation, which occur when extreme velocities in water mains cause attached biofilm to shear from interior pipe walls. This typically becomes a local rather than a systemwide water quality concern. Regular cleaning of water mains and control of water main velocities reduce the incidence of biofilm separation.

DISTRIBUTION SYSTEM VULNERABILITY

There are several distribution system characteristics that make it vulnerable to failures that impact public health.

As part of the workshop held for this project, the attendees were asked to identify the key issues that a utility should be managing. Appendix A summarizes the workshop details and findings. These characteristics can be defined as engineering features, operation and maintenance (O&M) practices, personnel issues, and factors beyond the utility's control. Utility-specific failures are briefly summarized in Table 2.1 with a brief discussion of issues for which water utilities have sole responsibility.

PREVENTION OF INFRASTRUCTURE FAILURES

The utility should have programs in place that are designed to prevent infrastructure failures. These range from initial design decisions made before construction begins to annual maintenance programs for validating that the municipal water infrastructure system is performing as expected.

Infrastructure System Design

Water utilities should maintain and utilize a documented set of design standards for all components used in their entire distribution system. These standards should, at a minimum, meet the standards from AWWA and NSF International, or other nationally and internationally recognized standards organizations such as the Canadian Standards Association, and so forth. These standards can also provide additional guidance for issues such as allowable pipe and fitting

Table 2.1
Water infrastructure characteristics impacting public health

	Engineering Features
Design criteria	Design criteria for water systems have traditionally focused on provision of sufficient water for firefighting and maximum daily customer demand, as well as reliability, by means of pipe looping, to supply water when a portion of the infrastructure is out of service. Design criteria must be given due consideration because of the expected longevity of distribution infrastructure—greater than 100 years for some pipes. Negative impact: When distribution infrastructure is oversized for the typical needs of the community served, there can be significant degradation of water quality. This is due to increasing water age within the infrastructure, which can be complicated by the presence of short-circuiting. Positive impact: These design criteria result in a robust system that is able to supply water during emergencies. Alternative paths for the water utility to maintain water supply or to secure alternate supplies to customers during major infrastructure failure must be considered and provided.
Construction	Construction of water mains requires careful monitoring to ensure that installed water mains remain free of defects and debris. Component materials should be inspected for elements that could be potentially harmful to human health; examples include lead in metals and solders, as well as various pipe lubricants used at joints. Components must meet AWWA and NSF International (NSF) standards. A utility should document their water main commissioning standards and have an inspection procedure for all infrastructure connected to the system. The utility should also specify acceptance and rejection standards. These should be used when private developers construct potable water infrastructure for privately constructed municipal developments and subdivisions that will ultimately be connected to municipal water systems.
Location	Water mains typically cross other buried utility lines such as electric, telephone, sewer, and others. In some cases, water mains run parallel to other buried service lines. This can make water mains especially vulnerable to contamination from sewer mains should there be insufficient separation of buried lines. Inaccurate mapping can lead to puncture of water lines by other utilities, such as electric or telephone, during construction. A process for communicating and reviewing plans of owners and operators of other buried utilities should be in place to identify potential conflicts. This should occur before construction begins.

(continued)

Table 2.1 (Continued)

Age	Today's design standards apply only to water mains currently in place. Some water utilities may have infrastructure components older than 100 years. Often this leaves a utility with infrastructure that is subject to corrosion failure, resulting in water disruption and/or water quality complaints (e.g., red water, black water, taste and odor complaints), as well as being a potential source of lead leaching into potable water. Other areas of the community may suffer from poor water circulation, resulting in increased water age and loss of disinfectant residuals. Proactive programs should be in place to monitor water quality in existing areas, and there should be functional O&M programs in place to respond to developing infrastructure issues. Utilities can develop "Nessie" curves showing the relative volumes of infrastructure that fall into the different age categories and develop renewal plans around the highest value assets. An example of a Nessie curve is illustrated in the AwwaRF report, *Financial and Economic Optimization of Water Main Replacement Programs* (Cromwell et al. 2001).
Accessibility	Every customer with direct connection to the distribution system has the potential to intentionally, or unintentionally, contaminate the municipal water supply by introduction of a foreign substance from a cross-connection and/or backsiphonage.
	O&M Programs
Frequency and types	A well-run utility should have a series of O&M practices that focus on maintaining water quality in the distribution infrastructure. These programs should consider reliability of valves and hydrants to operate effectively when required, as well as component structural integrity. Reservoir maintenance programs must be in place to ensure that stored water remains at the best possible quality. Examples of these types of programs are described later in this chapter.
Water quality monitoring program	A water quality monitoring program that is sufficiently sensitive to detect and discern seasonal water quality and quantity variations is an essential tool in reducing vulnerability of the water utility to infrastructure failure. Understanding how water quality can typically vary throughout the year can help proactively detect, identify, and solve water quality problems before they develop into water emergencies. Random sampling, coupled with rapid response and investigation of all customer concerns can prevent the potential for widespread contaminant dissemination.

(continued)

Table 2.1 (Continued)

Presence/absence of SCADA (supervisory control and data acquisition)	A SCADA system can be an effective tool for monitoring and controlling the flow of water in a distribution system. This tool is more common within a water treatment plant environment, but can also play an important role in water infrastructure operation and management. Typical uses include monitoring of pressures at key locations in the distribution system, monitoring of reservoir levels, and control of pumping and filling of reservoirs. These tools allow the water utility to continuously monitor their distribution infrastructure, as well as provide advance notification of infrastructure system upsets or failures through early-warning systems and alarms.
Personnel Issues	
Education	Staff should receive regular training in issues related to preventing failures, as well as those needed to ensure public health protection. Appropriate incident and/or emergency procedures should be developed and practiced so that coordinated utility responses to failure events are rapid, well managed, and effective.
Attitude	Utility managers should adopt an attitude of awareness that production and delivery of safe drinking water is a critical community public health requirement. This message should be emphasized frequently to all water utility staff.
Communication	Regular communication channels allow easy exchange of routine information and ensure appropriate contacts are established that will quickly facilitate the handling of potential or confirmed emergency situations. Communication protocols should be set up between the utility, license regulators, and local health authorities. Each party should be aware of their individual roles, responsibilities, and jurisdiction, as well as those of the other partners, to ensure safe drinking water and public health protection.
Factors Beyond the Utility's Control	
Natural disasters	These include unforeseen and unpredictable events such as tornadoes, hurricanes, floods, landslides, ice storms, and earthquakes. All are capable of causing significant infrastructure failures allowing contaminants to enter the water main; either at the location of the failure, or by causing low pressure in the surrounding distribution system that results in backsiphonage. More common natural events that a utility must be prepared to deal with are power failures, which impact pumping capabilities and hence, delivery pressures.

(continued)

Table 2.1 (Continued)

Human-caused disasters	Problems can result from human error such as puncturing an existing pipe, opening or closing the wrong valve, or accidental spill of chemicals. When these types of situations occur, the utility should maintain a culture of openness so that such events can be reviewed and learned from. Any procedural issues that have been identified should be used to revise future training programs and to implement necessary changes to the utility's standard operating procedures (SOPs). Utilities may also be faced with intentional harmful activities specifically designed to adversely impact the municipal water system. A vulnerability assessment can assist the utility in understanding how to manage these risks and develop necessary response plans.
Public fears/rumors	Water utilities may have to respond to public fears and rumors arising from a perceived infrastructure incident or an actual event. In these types of situations, the public must be assured of their significance in the community. It is important for the utility to have a good relationship with local media, local emergency responders, and any other agencies that must be involved. This can ensure that accurate information is presented and can help reduce public fears and avoid unnecessary panic.

materials, installation techniques, cathodic protection requirements, fire-flow standards, and pressure ranges for operation. System design should include consideration of reliability and water mains looping during periods of valve closure, as well as adequate water supply during staging of new developments.

From a water quality perspective, the number of dead ends and low-flow areas should be limited to prevent stagnant water; this includes the length of hydrant leads. A formal process to evaluate new materials and technologies should be in place that allows utility standards to be continually updated to comply with best available industry standards.

Infrastructure Renewal Programs

A renewal program can help reduce the number of unplanned failures and therefore limit opportunities for contaminants to enter the distribution system and adversely impact the community's health.

Notwithstanding the use of exceptional engineering design and construction standards, water utility infrastructure begins to age and deteriorate as soon as it is exposed to the surrounding environment. Rather than taking a reactive approach and dealing with upgrading and/or repairing existing water infrastructure on a failure-by-failure basis, the water utility should have a series of ongoing renewal programs to upgrade existing infrastructure in a proactive manner.

Typical utility renewal programs include water main replacements that are based on the frequency of main breaks. There are also programs that address the need for lining the interior of water mains in order to reduce water quality concerns caused by red water. Finally, many utilities have cathodic protection programs in place to slow the rate of external mains corrosion.

Infrastructure renewals can also be performed to bring existing infrastructure up to newer design standards that are capable of providing higher fire flows and that also support intensification of land use and increased customer water demand. In the United States, the recent adoption of the GASB 34 (the Governmental Accounting Standards Board, Statement 34, *Basic Financial Statements for State and Local Governments*; GASB 2003) regulations requires water utilities to focus on renewal of existing infrastructure as part of their financial accounting practices.

Preventive Maintenance Programs

Routine system maintenance should include programs intended for inspection and repair of all elements of the distribution infrastructure. These programs should have documented SOPs, as well as ongoing staff-training programs to ensure that these activities are completed efficiently while minimizing the impact to the customer. The following utility programs specifically impact water quality and, hence, public health.

- Water mains cleaning: All water utilities should have a program for routine cleaning of existing water mains. The type of cleaning technique required will vary based on distribution system conditions and delivered water quality. Standard methods of mains cleaning include conventional flushing, unidirectional flushing, pigging, and air scouring. Regular cleaning can reduce the impact of system failures by ensuring that biofilms remain at a minimum. In the case of unidirectional flushing, routine mains cleaning confirms that valves and hydrants can operate as intended and when required (e.g., during fires). For additional mains cleaning information, refer to a report published by AwwaRF in 2003 titled *Investigation of Pipe Cleaning Techniques* (Ellison et al. 2002).
- Hydrant and valve inspections: These inspections can be completed as part of a unidirectional flushing program; however, a utility should also maintain a separate program that reviews areas not flushed in a given year. Typically, hydrants should be inspected and operated a minimum of once per year. Some municipal fire insurance standards may require more frequent inspections. Regular operation of the fire hydrant is important to remove stagnant water in the hydrant lead. This water can enter the municipal distribution system as a result of backsiphonage during a mains pressure transient event. Valve inspection can occur less frequently; however, high-risk valves must be regularly operated to ensure they will function as required. Care must be taken when operating distribution valves to ensure that water supply to the affected neighborhood is not interrupted.
- Leak detection: Leak detection programs involve the regular use of inspections and acoustic listening devices to identify the location of leaks for repair purposes. Leak detection and repair is important to reduce water losses in the distribution system as well as to reduce the potential for catastrophic pipe failure due to washout of bedding materials supporting the leaking water main. In areas near sewer mains, water leakage can also lead to potential contamination.
- Transmission main inspections: Transmission mains, because of their larger size, will have the greatest impact when they fail. A utility should have regular inspection programs for potential cross-connections to transmission mains. Transmission valves should be inspected and operated to ensure that they will function as required.

Cathodic protection systems on transmission mains should be maintained to ensure that they are functioning properly.

Reservoir Management and Maintenance Programs

Water reservoirs may consist of standpipes where the water level "floats" on the distribution system pressure to buried ground-level reservoirs that require pumping to deliver water to the distribution system. In all configurations, there is the potential for water quality to degrade during storage. The degree of degradation can be limited through proper operation of reservoirs to ensure adequate reservoir circulation and turnover, and through regular maintenance programs to keep the reservoirs clean.

Daily or weekly reservoir storage plans are a good tool for monitoring reservoir usage levels and for managing water age. AwwaRF studies recommend either complete reservoir turnover every 3 to 5 days, or improved reservoir mixing so that water within reservoir is well mixed in less than the normal reservoir fill time (Grayman et al. 2000, Kirmeyer et al. 1999).

In the case of standpipes that float on the distribution system, pressure piping configuration should be designed to ensure that bands of stagnant water do not develop within the standpipe. This stagnant water may potentially be released to the distribution system during a hydraulic transient. For pumped reservoirs, the inlets and outlets should be designed to ensure adequate mixing of incoming fresh water with the oldest water in that system, which reduces the potential for stratification within that reservoir. Pumps can be operated to ensure that predetermined minimum turnover rates are maintained.

Maintenance programs should include regular inspections of all hatches, vents, and access points. Any signs of tampering must be thoroughly investigated. If possible, alarms should be added to all access locations and a defined plan of action developed in case of alarm activation. Limitations for the types of chemicals to be used on any grass in the immediate area of the reservoir or on the reservoir roof should be clearly posted.

Regular full cleaning and inspection of reservoirs should be planned for any reservoir in service. The frequency of cleaning can be determined by assessing the extent of sediment accumulation within the reservoir. Typical schedules are once every 5 years. As part of the cleaning regime, there should also be an inspection of all cracks, internal pipe corrosion, and roof integrity by means of a flood test. Caulking materials used in new reservoirs, or as part of a repair, can result in odors or low-level trace organics; additional reservoir flushing may be required before putting the reservoir back into service.

Some communities are limited in their ability to clean potable water reservoirs because of poor initial design and lack of redundancy, which prevent the reservoir from being taken out of service for cleaning. In these types of situations, the community should consider using remote cleaners that work in the reservoir without requiring the tank to be drained or a community cooperation program with temporary provisions for truck-hauled water in order to meet basic needs only.

Cross-Connection Control Programs

Cross-connection control is generally carried out with the use of a formal program that requires the installation of backflow-prevention devices on sites that are at greatest risk of contaminating the drinking water supply because of backsiphonage. A cross-connection control program may be required as part of the state or provincial plumbing code or municipal by-laws. Other regulatory agencies, such as departments of municipal affairs, public health, or environment, may have pertinent regulations that specify requirements and installation for backflow-prevention devices as well as for inspections. Responsibility for a cross-connection program may lie with the water utility or be the sole responsibility of the building owner. This must be clarified with local authorities. Backflow-prevention devices are usually tracked and inspected as part of a utility-based inventory having regularly scheduled inspection visits. Cross-connection control programs should include annual inspections of backflow devices to ensure proper operation; some utilities issue certificates of inspection. The extent of implementation of cross-connection control programs varies throughout North America.

When the installation and proper operation of backflow devices at high-risk locations are effective, the ability for rapid investigation of suspected backsiphonage events becomes a primary defense against intrusion of waterborne contamination into the distribution system. For this reason, all customer complaints should be considered seriously and investigated thoroughly. Customer complaint types and geographic locations should be mapped to determine whether patterns or trends are emerging.

Any suspected cross-connection incidents must be investigated quickly. Potential contaminants, and the extent of contamination, should be identified as soon possible; local customers should be warned to prevent possible health-based consequences. Alternative sources of potable water should be provided if needed. The investigation should continue until the problem is resolved. If the issue is an internal building problem, the owners should be informed of their responsibility to remedy the problem.

WATER QUALITY MONITORING

A routine water quality monitoring program should be set up to provide sufficient information on water quality to characterize the entire water service system. Monitoring should include water entering the distribution system, reservoirs, representative random sampling sites, dead-end and/or low-flow areas, frequent customer complaint areas, different pressure zones, transfer points to major bulk water customers, and high-risk sites such as hospitals, day-care facilities, seniors' homes, hotels, and so forth. Dividing the community into specific zones and sampling each zone regularly can be useful for ensuring that the program is continually monitoring averaged water quality levels for the entire community. Zone boundaries can be determined based on land use and/or population levels in each area to ensure equal representation.

Monitoring of water treatment plant effluent on a continuous or daily basis for chlorine residual, turbidity, color, coliforms or *Escherichia coli* bacteria, and other parameters of interest is very important and ensures that the potential for systemwide contamination is minimized. Routine source water monitoring is useful to identify periods of increased challenge levels for waterborne contaminants to enter the water treatment plant. During these periods, the utility should consider increasing the monitoring frequency and keep public health officials informed of variances from expected operating conditions.

All customer complaints or concerns should be investigated. If the complaint appears valid, samples should be collected for analyses of turbidity, chlorine residual, color, odor, volatile organics, pipe lubricant, iron, conductivity, pH, bacteria (i.e., coliforms, heterotrophic plate count [HPC], etc.), as well as any other analytical parameters necessary to determine the cause for complaint. The process should be sensitive enough to distinguish between issues with municipal mains and building plumbing issues. Customer complaints can provide an early warning of contamination of the municipal system. Once main break repairs have been completed, the affected area should be flushed and then sampled for turbidity, chlorine residual, and bacteria to confirm that the water quality has been restored to normal levels.

It is useful for the water utility to have an internal policy stating that water quality in the distribution system be maintained at higher standards than regulated license requirements. Regulatory requirements are the minimum requirements that a water utility must meet; the utility may wish to consider higher standards. Working to a higher standard usually prevents minor upsets from immediately becoming a regulatory violation and/or a public health concern. A well-maintained distribution system allows for early warning and detection of potential problems; security that a poorly maintained system cannot provide. Internal guidelines for distribution system quality, such as setting goals of turbidity to <1 ntu, a chlorine residual limit to double the regulated minimum level, and HPC warning levels to 10 or 50 cfu/mL, are useful tools for maintaining excellent system quality.

Water quality data can be mapped on geographic information systems (GISs) to provide information on systemwide problems or to detect clusters of potentially poor water quality. Used in conjunction with a hydraulic model, this review can help develop action plans to improve water quality in a targeted area of the municipality. Tracking water quality data over time in particular zones or at key monitoring locations can help determine whether there are seasonal issues, water temperature concerns, or deterioration of quality in a specific area over time. Water quality data should be shared freely and in a timely manner with all other agencies involved with safe drinking water in the community. Conversely, health agencies should make available to the water utility any data on suspected or confirmed waterborne disease occurrence in the community; this information can be used to track trends and determine whether links exist between the community's drinking water and particular health indicators.

MANAGEMENT OF FAILURES

Even with all-encompassing failure prevention and water quality monitoring programs, infrastructure failures may occur within the distribution system. This section will review the most common types of infrastructure failures and some techniques that the utility can use to mitigate the impacts of these events.

Structural Failures

The most common type of structural failure is a main break. These usually occur in older metal pipes and are caused by external and/or internal corrosion of the pipe wall. Structural failures can also occur from natural disasters that provoke ground shifting, causing the water main to break. Human-caused structure failures may arise from puncturing a water main during adjacent construction activities.

In most cases, the impact of the structural failure will be a localized event and limited to the surrounding area, that is, a few municipal blocks. In the case of the structural failure of a major transmission main, the impact can be widespread and long lasting, requiring alternate water supplies to be provided to large portions of the system for an extended period of time.

When there is a structural failure, there is an accompanying loss of pressure in the distribution system. This can lead to backsiphonage by means of cross-connections from attached services. The failure may also introduce soil or other contaminants from the surrounding area into the water main. This is particularly critical if the water main is close to an adjacent leaking sewer line or if the sewer line was ruptured by the same event as the water main failure.

Rapid response to structural failures is essential for controlling the loss of water and the degree of washout damage to surrounding roads and utilities. Local customers should be notified quickly and the area of the failure isolated. Appropriate valves should be closed as soon as possible to restrict the spread of possible waterborne contamination. If numerous customers are impacted by the failure, the target should be to return the water main to service as soon as possible. To mitigate the impact of structural failures on critical water mains, a utility should conduct a critical mains assessment. Any water mains that are the sole supply source to an area, or that have an unusual material or size constraint, should have specific response plans in order to minimize the length of time that these mains remain out of service.

Utilities should maintain a list of critical customers who may be more severely impacted because of the temporary loss of water during a failure event. Typically, this list would include hospitals, schools, day-care facilities, nursing homes, and possibly those on home dialysis machines. This list would enable the utility to provide extra support to these customers. There should be ongoing, regular communication with the public health department during these events.

Although a utility will be in a reactive mode when responding to infrastructure failures, performance standards can still be developed and monitored to improve the quality of the response to these events. Typical monitoring parameters that can be measured are response time to isolate the failure, length of time to repair, and cost of repair.

Power Outages

Power outages can be significant problems for water utilities that rely on pumps to pressurize their distribution system. Extensive outages (for hours or days) can result in loss of treatment capabilities as well as loss of pump usage to maintain system pressures. Short-term outages, 15 minutes or less, can also be an issue because the loss of pumping ability can create a hydraulic pressure transient. This can result in low pressures and, in some cases, negative pressures in the distribution system, thus increasing the potential for a backsiphonage event.

A utility can minimize the impact of power outages by having multiple pump stations for each area of their system, each pump station being energized from a different power substation. For critical pumping stations, the utility may want to consider providing two electrical power feeds to a single location, each from a different power substation and each with automatic switching gear to allow for seamless transition should one source fail.

Standby generators can also be installed; however, these will require regular maintenance to ensure that they will function when required. Noise complaints may become an issue if the supply location is in a residential neighborhood. Fuel supplies and fuel storage may also become issues, depending on where the standby generator is stored.

Electrical brown-outs have become an issue in the past few years in areas where electrical generation capacity cannot meet customer demands. In these cases, the water utility should be communicating with the local power utility to ensure that water utility pumping stations are considered critical infrastructure and therefore not included in the brown-out rotation.

A hydraulic model is an excellent tool for assessing the impact of power outages, as well as for providing water utility operations staff the opportunity to practice different failure scenarios before they occur.

Water Quality Contamination Response

Water quality contamination will usually be noticed as the result of a series of customer complaints about visible changes in delivered quality; by a known failure such as backflow, spill, main break, pressure drop, and so forth; or through a regular utility water monitoring program that tracks changes in routine parameters such as chlorine residual, turbidity, odor, or HPC.

Conductivity and pH can be quick, easily measurable indicators of extensive inorganic contamination, while ultraviolet (UV) absorbance scans or UV light at 254 nm (UV_{254}) readings can help determine organic contamination. Color and iron monitoring may be required in red-water complaints or rust-problem areas of the distribution system. Odor testing or trace volatile organic monitoring can identify the presence of contaminants such as solvents; benzene, toluene, ethylbenzene, and xylenes; or sewage contamination. Most of these tests can be performed relatively quickly, often within a few hours. If preliminary testing indicates a problem, more extensive testing may be required. Bacteriological testing, including tests for total and/or fecal coliforms, *E. coli*, and HPC, is advised in any suspected contamination event to ensure control of disease-causing organisms. Results for bacteriological tests take 24 hours or more, so it may be necessary to provide alternate water supplies or bottled water to impacted residents until the issue is resolved.

In any case, monitoring and analyses should continue until the contaminant is identified with reasonable certainty. Alternative potable water supplies should be available to affected consumers until the distribution system failure can be remedied. Water distribution hydraulic models can be utilized to determine contaminant propagation within the distribution system, and can be used to develop a flushing program to clean the system as quickly as possible. Because, in most cases, the response team is looking for the extent of contaminant propagation as opposed to the extent of degradation during travel, it is not necessary to have a water quality model to complete this analysis. The toxicity of the contaminant and the extent of the contamination must be determined rapidly so as to identify the number of affected customers as well as an estimate of the magnitude and potential short- and long-term impacts of the event. Appropriate resources and communication can then be applied to the problem. This usually involves multiple partners in the community, but especially the public health department at the local level, and possibly also at the provincial/state and/or federal level, depending on the severity of the situation.

The health surveillance system is the last, and very important, monitoring step for the detection of cases of illness as a result of an infrastructure failure. A passive health surveillance system will usually detect significant clusters of illness, indicating systemwide contamination or localized contamination provided enough cases are reported to the local health authority. Chapter 3 discusses the roles and responsibilities of public health agencies and the mechanisms of active and passive health surveillance systems. If the incubation time for the disease is relatively long, then the community health effects may not appear for up to 4 weeks after the infrastructure failure event. The actual contaminant in the municipal water supply may never be detected, depending on the length of the incubation period coupled with the number of days of water storage the utility has within its system. Often, epidemiological association and the strength of an association between the disease cluster and potential community sources may be the only tool available to link community morbidity with the infrastructure failure event. The passive health surveillance system may not identify very small, localized infrastructure failure events because of the relatively low level of contamination and the limited number of customers impacted.

Emergency Plans

Every utility should have emergency plans set up in advance for all situations that may have serious community consequences, such as floods, chemical spills, terrorist threats, tornadoes, hurricanes, major power outages, ice storms, and earthquakes. Responsibilities, procedures, and tasks should be written in advance, and staff should be well trained in responding to such events by routine practice involving mock emergencies or tabletop exercises. Chapter 4 discusses some of the steps required to build the necessary teams for handling community emergencies and provides guidelines for joint training.

The utility emergency plan should clearly define the following elements:

- Authority and jurisdiction for invoking and coordinating emergency procedures
- Definitions and characterization of risks
- Identification of key personnel and their roles
- List of specific responses for identified risks
- A developed boil-water order or no-water advisory, as well as protocols for other types of emergency contamination events
- A list of communication procedures for the media
- A list of other organizations that may need to be involved, such as the local health department, environmental regulator, fire department, police department, chemical spill response teams, etc.
- A list of available emergency equipment such as generators and spill kits
- A list of alternative potable water supplies such as bottled water and water haulers
- Any mutual aid agreements needed for severe situations

Followup procedures after a disaster may be extensive, expensive, and require nonroutine methods and/or procedures. Often, multiagency and/or multiple municipality cooperation and collaboration is also required. Ideally, these agreements should be in place before an incident and/or emergency occurs. Special oversight groups and extra staff may be required for several weeks to several months after the event.

Emergency plans should be readily available at designated control centers in a nonelectronic format; this is essential if power to the emergency center is disrupted.

SYSTEMS AND TOOLS

The following sections briefly describe some of the systems and tools that a water utility can use to manage infrastructure failures and ensure continued protection of the public's health.

Geographic Information System

A GIS is a mapping and analysis tool that allows water utilities to display the water infrastructure as well as other information such as land use, customer types, road types, and soil conditions. This data can be analyzed to prioritize areas for water main renewal, water quality sampling, and developing appropriate response plans to widespread and/or localized infrastructure failure events.

Over the past few years, GIS tools have become much more affordable, as well as more user friendly. The availability of open architecture standards and standard user interfaces have resulted in more GIS tools available at an individual desktop computer that facilitates rapid analysis with limited specialized training required. In many communities, base maps are commercially available, thus allowing the utility to add community-specific information in order to fully utilize the GIS.

Maintenance Management System

A maintenance management system can be as simple as a series of file cabinets and a large whiteboard for scheduling activities in a small community. It can also be as complicated as a computer system installed on wide area network with mobile field computing to capture and dispatch activities from the field staff. In both cases, the basic principles are the same: the coordinated ability for scheduling and monitoring all infrastructure maintenance activities. A proactive infrastructure maintenance program, able to schedule and track activities for the entire system, becomes the foundation for an infrastructure renewal program. All reactive maintenance should be tracked in the same system to allow development of proactive programs, especially if evidence indicates that a new trend is developing or that a particular type of infrastructure appears to be failing at an accelerated rate.

Hydraulic Model

Hydraulic models of water utilities are uncommon in the water industry, yet they were often built to assist with capital expenditure planning. If a water utility's engineering department has a hydraulic model, consideration should be given to providing field crews access to the model to assist with day-to-day operational planning. This would involve modeling planned shutdowns, then validating that expected results were achieved. The hydraulic model is also an excellent tool for practicing responses to major system failures. A significant, ongoing benefit to the water utility in having operations staff regularly use these tools is the continual validation of hydraulic model accuracy. It becomes a year-round calibration process rather than a periodic effort. There is also a greater potential that improperly closed valves will be located because field validation results will not match hydraulic model predictions.

Laboratory Information Management System

A laboratory information management system (LIMS) collects all of the utility's water quality data, together with pertinent information on sampling date, location, reason(s) for sampling, (re)sampling requirements, and so forth. The LIMS should be easily searchable and be able to quickly produce reports on demand. The LIMS can be as simple as several spreadsheets, but a full electronic database system would allow the utility to link results of sampling with a GIS and a hydraulic model to more comprehensively analyze the source(s) of change in delivered water quality. The GIS linkage can also be used to ensure that collection of random samples truly represents the entire distribution system, as well as the different types of customers serviced.

Intranet/Internet

Intranets and the Internet are effective tools for communicating with utility staff, other agencies, and the public about issues impacting the water utility and, possibly, the quality and quantity of delivered water.

An intranet can be used to access SOPs, allowing the most up-to-date version to be accessible to all staff at all times. A backup hard copy should also be maintained in a designated area in the event of a computer or power outage. The Internet can be set up for communication with the public about ongoing and/or emergency events. Most customers now expect to be able to go to a company Web site and quickly find the information they are looking for.

For communication with other responding agencies, it is possible to set up a private Internet location requiring a password-protected logon to access the information. This is an effective tool for sharing information. In some communities, this technique is being used to submit information to a regulator on water quality sample results.

Customer Information System

Customer information systems traditionally have been limited to customer billing systems, where the primary focus has been to have the correct monthly water consumption and the correct address to send the bill. These systems can, however, be a great source of information for building the consumption patterns in a hydraulic model and for classifying customers in order to determine the impact of distribution failure, and for developing contingency response plans in the event of a major failure.

CHAPTER 3
ROLES AND RESPONSIBILITIES OF PUBLIC HEALTH AGENCIES

Nelson Fok and James A. Talbot

INTRODUCTION

The goals of public health improvement and protection are to reduce the amounts of discomfort and disability, disease, and premature death in the population (Last 1998). In the case of water distribution system infrastructure failure, as with other disasters, the duties of public health agencies are primarily directed toward prevention or mitigation. Primary prevention refers to efforts made to prevent failure from occurring. Secondary prevention includes those efforts made to mitigate the event by advance preparation, early recognition, and prompt action to reduce community health consequences (Sidel et al. 1992).

Failure can be classified into two categories: sudden or insidious, and continuing (Sidel et al. 1992). Estimating health risks from infrastructure failure mostly depends on the use of epidemiological techniques to describe, estimate, and measure adverse health effects and the factors that contribute to them (Noji 1997). Epidemiological techniques are used to rapidly define the nature and extent of health problems, identify groups at particular risk, optimize relief operations, and recommend ways of decreasing the consequences of future emergencies and/or disasters (Glass and Noji 1992). Planning for prevention of emergencies and mitigating their effects, if prevention is unsuccessful, has been divided into three phases: preimpact, impact, and postimpact (Binder and Sanderson 1987). Each phase may require different methods for collecting, analyzing, and disseminating the data (Wetterhall and Noji 1997).

This chapter is divided into the following sections:

- Differences in perception of risk assessment
- Water distribution system failure conditions and resulting hazards
- Primary prevention (prevention of the occurrence of failure)
- Secondary prevention (mitigation of the consequences of failure)
- Impact surveillance programs, including monitoring and surveillance tools that may be of use in detecting and measuring the impact of a failure
- Incident response strategies, including those to effectively move from incident to corrective action
- Postimpact surveillance programs, including surveillance to ensure that corrective actions were effective, and planning for future preventive strategies

DIFFERENCES IN PERCEPTION OF RISK ASSESSMENT

Water delivered to the community is an essential element for health. Historically, the partnership between public health authorities and the water industry has been crucial to improvements in health and reduction in morbidity and mortality. Mutual understanding and cooperation has been central to this success. The workshop held in preparation for this report revealed misunderstandings between the water distribution system operators and public health officials that have the potential to be divisive if not addressed.

Responsibility for Public Health Versus Product Liability

Public health officials are ultimately responsible for protecting the health of the community. Water distribution system failures can manifest as disease in the population served; public health agencies have a statutory and jurisdictional interest in being informed of these threats. The water distribution system operator has the responsibility to monitor, understand, investigate, and act to preserve water quality and to inform public health regulators when potential or actual harmful exposures may or do occur. Public health agencies have the ultimate regulatory responsibility for protecting the health of the community. If an infrastructure failure has the potential to affect health, it becomes the joint responsibility of the public health department and the water utility to assess the risk and then to determine all necessary actions to protect the community.

Risk Analysis Versus Risk Assessment

Historically, risk assessment has been confused with analyzing the level of risk and controlling the risk. Risk estimates were often affected by the cost of managing the risk (Gochfeld 1992). For this reason, risk analysis was introduced as a method of estimating risk, independent of management or economic concerns. A risk analysis approach is more comfortable for public health officials, in part because the mandate of those officials is to protect health. This approach is free from requirements such as that of water distribution system operators to provide a safe product at an affordable cost. Therefore, even if a problem could be prohibitively expensive to fix, public health officials and/or regulators must be informed, and, once again, public health officials have the ultimate responsibility for assessing risk and ensuring that corrective risk management occurs. The risk management options available to public health officials are broader and, in some situations, likely to be more effective or safer than those available to water distribution system operators acting alone.

Acting Without Solid Evidence

Public health agencies rely on the precautionary principle when evaluating potential threats to the health of the community. The Wingspread principle states, "When an activity raises threats of harm to human health or the environment, precautionary measures should be taken even if some cause-and-effect relationships are not fully established scientifically" (Science and Environmental Health Network et al. 1998).

Adoption of the precautionary principle by public heath officials creates a potential division with water distribution system operators who usually operate within a framework of risk assessment and management. Risk assessment relies on accurate information to properly perform hazard identification, hazard assessment, exposure identification, exposure assessment, and risk characterization (Gochfeld 1992). Applying this logical and sound scientific research approach can be problematic in the real world of evaluating the public health consequences of distribution system problems and/or failures (see Table 3.1).

Gathering accurate information to establish hazards, exposures, and risks can be difficult in actual communities. As an example, elementary knowledge such as the location of vulnerable populations, the health effect in those populations, or the dose magnitude and duration to which they have been exposed are often unknown. In the real world, monitoring equipment breaks down, employees may falsify information, and surrogate tests may fail to accurately predict exposure.

Table 3.1
Different interests of public health officials and environmental engineers

Axis of Interest for Risk Assessment

		Harm	
		Yes	No
Knowledge	Yes	Known harm	Known safe
	No	Unknown harm	Unknown safe

Axis of Interest for Precautionary Principle

		Harm	
		Yes	No
Knowledge	Yes	Known harm	Known safe
	No	Unknown harm	Unknown safe

Notification of infrastructure events, such as backsiphonage or main breaks, to outside agencies may not occur to the water utility if the apparent significance and potential public health consequences are not considered to be important. If action is taken only after significant scientific proof of harm has occurred, the opportunities to prevent exposure or mitigate disease are seriously compromised. The precautionary principle dictates that failure to possess accurate information is not sufficient to justify inaction on the part of public health. Therefore, having knowledge of an infrastructure event, even without conclusive evidence of community harm immediately available, there remains the need to preemptively act in the best interests of the community.

WATER DISTRIBUTION SYSTEM FAILURE CONDITIONS AND RESULTING HAZARDS

As further described in chapter 2 and appendix A, workshop participants identified three general categories of water distribution system failure conditions that impact the quality of the public's health. These were structural failures, cross-connections, and intramural hazards.

Structural failures, such as main breaks or reservoir failure, can result in risk due to the entrainment of biological or chemical hazards found in the immediate environment surrounding the break or within the actual storage reservoir. By far the most common hazard mentioned in the workshop was leakage from sewer pipes contained in the same trench as the water pipe, or in close proximity. This type of hazard and other exposures that result in structural failures can only be assessed with knowledge of the hazards found in the area affected.

Cross-connections occur most commonly through mismatches between higher pressures in customer properties, resulting in net flow back to the lower-pressure water distribution system mains. The exposures that result can only be predicted with knowledge of the hazards introduced from the customer's establishment. Cross-connections usually result from accidental pressure differentials but, theoretically, could also be the result of deliberately created differentials.

Intramural hazards, such as biofilm separation, trihalomethane (THM) formation, leaching of lead, and *Cryptococcus* in treated water reservoirs (Doggett 2000), result from processes occurring within the distribution system. Biofilm separations typically result from boundary shear at the pipe–water interface due to extreme velocities, such as those that result from fire-suppression activities. The potential for public health risk from these events requires knowledge of the nature and extent of the event, as well as the type, concentration, and exposure routes of the hazardous material within the biofilm sheared. Therefore, whether potentially pathogenic and/or harmful material is contained within or on the biofilm, as well as the effectiveness of biofilm removal routine maintenance, must be determined. THM formation is usually dependent on chlorine residuals acting on organic carbon, including that found on biofilms. Lead is an example of a hazardous material that may be leached from the pipes, joint compounds (e.g., solder), reservoir liners, and household fixtures.

Primary Prevention (Prevention of Failure Occurrence)

The workshop participants routinely used the environmental risk assessment approach; this section will be organized with the same structure. The components include hazard identification, hazard assessment, exposure identification, exposure assessment, and risk characterization (see Table 3.2).

From Table 3.2 it is possible to identify the information that would be necessary to estimate the health risk from infrastructure failure. As an example, a community that wished to assess its risk from lead pipes in the distribution system would need to know how many miles of pipes were in the system, what concentration of lead was found in the water, what the community levels of lead were in the regions serviced by those pipes, and which especially susceptible populations were being exposed (e.g., children). Additional information would be required to judge the success of control measures (see Table 3.3). In the foregoing example, if the decision was made to replace the lead pipes, then information would need to be collected to show the replacement rate and, perhaps, declining population lead levels in the geographic areas where replacement had occurred.

Secondary Prevention (Mitigation of Failure Consequences)

Mitigating the consequences of a distribution system failure requires that the failure be recognized (surveillance) and that planned activities occur (incident response) to reduce the consequences of infrastructure failure, particularly on human health. The most important time to plan and test surveillance and incident response is in the preimpact phase (see Table 3.4).

Waterborne disease outbreaks can be averted through proper water distribution system monitoring and effective communication. This works best if there is a preexisting, functional, working relationship between the utilities and public health agencies. This may include setting up critical control points throughout the distribution system and defining monitoring parameters. Failure to meet monitoring criteria should result in notification to the appropriate public health authorities of incidents of infrastructure failure. It then becomes the responsibility of the public health agencies to work with the utilities to determine the likelihood of a potential waterborne

Table 3.2
Primary prevention*

Situation	Hazard: Identification	Hazard: Assessment	Exposure: Identification	Exposure: Assessment	Risk: Characterization
Main break	• Microbiologically contaminated soil • Sewer line in trench	Do the microbiologic agents survive? Are concentrations sufficient to cause disease?	• Community water testing. Increase in gastroenteritis/diarrhea. • Increase in pathogens isolated from stools of cases in community. Increase in emergency room visits, doctor's office visits.	• Exposure mechanisms: inhalation, ingestion, cutaneous • Concentration • Amount consumed	Quantitative estimate of exposure level that leads to particular excess response (dose response)
	• Chemically contaminated soil • Leak from gasoline filling station reservoir	Do the chemical agents persist? Are concentrations (likely to be entrained) after a break sufficient to cause disease?	• Community water testing. Customer complaints of odor or taste. Increase in acute or chronic toxicity cases. If carcinogen, increase in cancer cases.	• Exposure mechanism: inhalation, ingestion, cutaneous • Concentration • Amount consumed	Quantitative estimate of exposure level that leads to particular excess response (dose response)
Cross-connection	• Microbiological agents introduced into water	Do the microbiologic agents survive? Are concentrations sufficient to cause disease?	• Community water testing. Increase in gastroenteritis/diarrhea. • Increase in pathogens isolated from stools of cases in community. Increase in emergency room visits, doctor's office visits.	• Exposure mechanism: inhalation, ingestion, cutaneous • Concentration • Amount consumed	Quantitative estimate of exposure level that leads to particular excess response (dose response)
	• Chemical agents introduced into water	Do the chemical agents persist? Are achievable concentrations sufficient to cause disease?	• Community water testing. Customer complaints of odor or taste. Increase in acute or chronic levels in population served or of toxicity cases. If carcinogen, increase in cancer cases.	• Exposure mechanism: inhalation, ingestion, cutaneous • Concentration • Amount consumed	Quantitative estimate of exposure level that leads to particular excess response (dose response)
Intramural hazard	• Pipe or connection material introduced into water	Is the material found in the water or is there evidence of pipe or connection erosion?	• Community water testing. Increase in acute or chronic levels in population served or of toxicity cases.	• Exposure mechanism: inhalation, ingestion, cutaneous • Concentration • Amount consumed	Quantitative estimate of exposure level that leads to particular excess response (dose response)
	• Material within distribution system water represents or creates hazard	Is the material found in the water or within the distribution system?	• Customer complaints. Community water testing. Increase in acute or chronic levels in population served or of toxicity cases.	• Exposure mechanism: inhalation, ingestion, cutaneous • Concentration • Amount consumed	Quantitative estimate of exposure level that leads to particular excess response (dose response)

Source: Adapted from Omenn et al. 1996.

* Identification: Can the hazard or exposure cause an adverse effect? Assessment: Does or will the hazard or exposure occur? Characterization: What is the relationship between the potential exposures and the incidence of adverse effects?

Table 3.3
Primary prevention and control

Situation	Hazard	Risk control: Questions	Risk control: Measures
Main break	Microbiological or chemically contaminated soil	Where is the contaminated soil? Do the agents persist? Which consumers would be exposed if a break occurred? How susceptible would they be? Are the concentrations (likely to be entrained) after a break sufficient to cause disease?	Replace pipes with less failure-prone construction. Remove contaminated soil. Change current construction methods so sewage pipe is not located in water pipe trench. Improve construction of sewage lines. Relocate pipes from dangerous areas.
Cross-connection	Microbiological or chemical agents introduced into water	Where are industrial and other locations likely to use or generate microbiological or chemical agents? Are the agents harmful? Is acute or chronic exposure harmful? Which consumers would be exposed to a cross-connection? How susceptible would they be? Are the concentrations, after a cross-connection, sufficient to cause disease?	Ensure that all known cross-connections are eliminated. Introduce backflow preventers into all situations that represent a potential risk.
Intramural hazard	Pipe or connection material introduced into water. Material within distribution system water represents or creates hazard.	What pipe and connection materials exist in the distribution system? What materials found or created in the distribution system exist in the distribution systems? How much material exists, and where are the materials located? Which components are potential risks? Is acute or chronic exposure harmful? Which consumers would be exposed and which would be susceptible? Are the concentrations sufficient to cause disease?	Replace pipe or connection material with healthier material. Introduce measures to reduce formation of THMs. Implement a regular maintenance program to reduce biofilm formation.

disease outbreak and to advise the community of the proper precautionary practices, including the issuance of a boil-water advisory or the provision of alternative water supplies.

Comprehensive emergency planning requires planning the response to an infrastructure failure, including delivery of adequate quality and quantity of potable water, before an event occurs. Public health agencies must be an integral part of every municipal emergency planning team, with representation at the control center, as well as being part of the media communication team. Chapter 4 describes in greater detail the roles of the various agencies in emergency planning

Table 3.4
Preimpact actions

1. Assess the level of preparedness of public health, environment, and utility staff.
2. Assess the capability of the surveillance systems.
3. Promote and educate public health, environment, and utility staff in the use of risk-reduction strategies.
4. Train utility personnel and public health agencies in risk-response activities and responsibilities.

Source: Adapted from Noji 1992.

and plan execution. During emergency planning for infrastructure failure, the following issues must be determined:

- Availability of an adequate, alternate safe drinking water supply from outside the affected area
- Adequacy of the alternate drinking water supply transportation system
- Adequacy of the dispensing system

Surveillance

The probability of detecting a waterborne disease outbreak depends on adequate knowledge of potential hazards, case definitions appropriate to those hazards, surveillance mechanisms to detect unusual clustering or trends of cases, and resources adequate to meet these needs. Detection of unusual clustering or trends of cases is improved by prior knowledge that water distribution problems were identified in particular locations (e.g., downstream of a main break) or at specific times (e.g., after a fire). This improves the probability of distinguishing between background case incidence rates and those directly related to the infrastructure event.

Long-term, low-level exposures that result in disease are much more difficult to detect. Therefore, in these types of situations, specific studies that compare normal disease incidence rates in specific affected populations may need to be compared to control populations lacking the long-term exposure. Rapid recognition of an outbreak and timely investigation greatly increase the likelihood of determining the cause. Without effective surveillance systems coordinated with water distribution system information, it is unlikely that an outbreak could be identified in its early stage.

Prior planning is necessary for proper outbreak management. The agencies and the allied professionals (physicians and laboratories) that must be involved should be identified for agreement concerning roles, responsibilities, and duties.

Active Surveillance

Disease rates are routinely monitored in day-care centers, hospitals, long-term care facilities, schools, correctional facilities, and religious institutions. Other databases of significant interest include hospital emergency room visits, family physician visits, health information hotline calls, and the sale of over-the-counter antidiarrheal medications. There is no "gold standard" that is able to preemptively predict the onset of a communitywide disease outbreak, irrespective of the cause.

Passive Surveillance

Notifiable diseases that are reported to public health agencies should be routinely analyzed to determine the background, as well as to benchmark, the number of enteric disease cases within the community. With sufficient data, it is possible to provide information on the most-likely causes, to project expected cases per month, and to predict seasonal variation as well as other factors that influence community enteric rates. Fluctuation in these rates can provide an early warning that there may be a potential outbreak of enteric disease within the community.

IMPACT SURVEILLANCE PROGRAMS AND INCIDENT RESPONSE STRATEGIES

When a potential waterborne disease outbreak has been identified, public health agencies are responsible for further investigation. These investigations are conducted primarily to ascertain the cause and to determine the size and nature of the outbreak. This is necessary to implement control measures that reduce the number of exposure cases. Skillful communication of the investigation and its findings to the public, primarily through the news media, is also necessary. The transmission of messages must be through a credible source to ease public concerns and fear, and must also be clear and precise to alleviate rumors (see chapter 4). In addition to informing the public about the progress of the investigation, useful strategies to reduce an individual's risk of contracting waterborne disease should also be communicated.

After protecting the public from exposure to chemical and microbiological pathogens in the water supply, the primary function for public health agencies during an infrastructure failure is to ensure that there is an adequate supply of water for essential functions. Public health agency support of essential functions is dependent on the severity of failure.

Failure may be limited to inadequate or unsafe delivery of water to a defined number of neighborhoods or communities by, for example, cross-connection and backsiphonage, sink-holes, leaking underground gasoline storage tanks and contaminated soil, main breaks and distribution line breakage, power loss confined to certain sectors, sabotage, or terrorism. In these cases, the basic infrastructure remains intact. Although the delivery of safe water is restricted or interrupted, the primary function for public health agencies is risk communication, implementation of risk-reduction strategies such as disease prevention by issuing boil-water advisories or provision of alternative safe water supplies, facilitating population removal if the former is not possible, or ensuring adequate fire suppression cooperation with distribution system providers to allow a reestablishment or redirection of water to affected communities.

Failure may involve inadequate or unsafe water delivery to the whole community because of complete power failure, extreme weather conditions such as an earthquake, hurricane, ice storm, flood, tidal wave, in addition to any potential for sabotage or terrorism. While the public health functions for this event are identical to those listed previously, the supplies, including potable water, necessary to support essential functions must originate from outside the community.

In severe events that lead to community water distribution system infrastructure failure, the first 48 hours are usually chaotic and, in many cases, are critical for survival. There must be an adequate supply of water for personal hydration, essential food services, emergency crews, fire suppression, and emergency services. Public health agencies must work closely with water utilities and other agencies to secure sufficient potable water for these uses. Temporary water supplies may be acquired from swimming pools, bottled water plants, ice-packaging plants, and water-vending machines. These facilities or operations have basic water treatment devices onsite and may have the storage capacity to provide emergency potable water for short periods of time. Public health agencies may also want to consider working with camping stores and the military to coordinate the use and distribution of point-of-use water treatment devices.

Failure may also compromise liquid waste, sewage disposal, and personal hygiene. This can result in an increase in enteric, person-to-person, and vermin transmission of disease. These considerations may intensify the need to set up temporary emergency shelters in schools, sports facilities, or community halls. Such centers have the advantage of providing potable water, waste disposal, and adequate food services in centralized locations. The establishment and maintenance of basic public health needs in these centers is the responsibility of community public health regulatory agencies. Often, the regulatory powers of the Medical Officer of Health in public health emergencies, as defined by provincial and/or state legislation, may be invoked to establish and supply these community crisis centers.

If movement to centralized community locations is not possible, then the public may need instructions for the safe preparation of infant formula and food, and provisions for general sanitation and personal hygiene. The public may also require advice on water conservation.

Safe Water Delivery

The delivery of safe water during infrastructure failure should meet the following objectives:

1. Ensure that all water to be used for human consumption is tested and determined to be safe before distribution.
2. Approve and inspect water haulers and water-dispensing stations.
3. Ensure a safe and adequate supply of water for infant formula and food preparation.
4. Ensure that a safe and adequate supply of water can be delivered for essential services such as those in emergency centers and hospitals.
5. Ensure that a safe and adequate supply of water can be delivered to vulnerable or immobile populations (e.g., newborns, the elderly, the disabled, and prisoners).
6. Ensure that a safe and adequate supply of water can be delivered to healthy individuals in the community.
7. Inform the public of procedures and locations to procure safe water and how to store it safely.
8. Provide information to the public about using point-of-use devices and emergency water sources such as water heaters.

The provision of adequate water supply for fire suppression is a joint responsibility between fire department officials, the water utility, and the public health regulatory agency.

POSTIMPACT SURVEILLANCE PROGRAMS AND FUTURE PLANNING

Postimpact surveillance systems must be customized to the specific incident that has occurred. In general, the critical point is that the management of the infrastructure failure should be measured against a previously agreed-upon set of objectives that specify the intended outcomes. The surveillance information is then used to examine the actual impact as compared to the desired management objectives. Specific and measurable objectives can be established for each category of distribution infrastructure failure, such as specifying the minimum amount of time needed until a specified number of community members have access to safe water supplies for consumption (Sidel et al. 1992).

If objectives are not met, information must then be collected to identify the reason for the discrepancy and necessary corrective actions implemented for future preimpact planning exercises and strategies.

CHAPTER 4
DEVELOPING AND MAINTAINING INTERAGENCY COMMUNICATION AND COLLABORATION

Janet Mainiero and Karen M.E. Emde

INTRODUCTION

Teams. Teamwork. Two simple words, yet subject to many different interpretations, applications, and methods of execution. For many, the word *team* may invoke the image of a sports or other competition; one team wins, therefore the other must lose. The competition continues with a single champion worthy of claiming the final prize among a battlefield of losers.

Safe drinking water teams are not a homogeneous blend of faceless individuals or agencies, but rather are composed of environmental and public health regulators, the water utility, emergency responders, engineers, laboratory scientists, municipalities, and the community as a whole. Building and maintaining a safe drinking water team doesn't happen by chance or luck. It requires long-term individual and corporate dedication to common goals, long-term commitments, and collaboration; it requires frank, open, and lively discussions; a willingness to listen to each other; and, finally, an unwavering desire to continue to work together. The true winner becomes safe drinking water for the community; there are no individual winners or losers. There is a common cause; there is a common purpose: safe drinking water for the entire community.

All too often, key players on the safe drinking water team first meet during a crisis. Tensions and emotions run high; regulators try to find and punish the violators; there are cases of waterborne illness and perhaps even death in the affected community; water utilities become defensive. The press carefully scrutinizes the situation for every newsworthy emotion, fact, rumor, fear, and, finally, every casualty. Recovery becomes a seemingly endless, tortuous road through the rumors, the fears, the facts, the misinformation, the regulatory inquiries and sanctions, and the suspicions. If the community safe drinking water team meets for the first time during an emergency, everyone ultimately loses.

The process of forming the team must begin long before the infrastructure failure, the emergency, or the incident ever happens. Forming a local safe water team and, more importantly, participating on this team, is neither instinctive nor intuitive. It requires a road map: where to start; what to do next; whom to consult and/or involve; how to train; how to change the plan; what to do when there appears to be a fork in the road; and, finally, how to best act and respond to incidents and emergencies as they occur. These events will surely arise, and usually during the periods of greatest community need, coupled with the least-available resources.

This chapter contains the framework and guidelines for pulling together a diverse and flexible team that will function as a unit during a normal situation or an emergency event. Suggestions will be offered on how to build trust, improve understanding, and encourage loyalty among members of the team. This chapter will clearly outline the key steps that are necessary to build and maintain a safe drinking water team. This process begins by starting to build your safe water team, which involves the following issues:

- Team building: roles of water utility, public health agencies, first responders, and other involved agencies

- Communications management:
  - Pre-event planning (normal situation)
  - Heightened alert situation
  - Emergency situation

To respond to and survive an emergency situation, two things must happen. First, the water utility, public health agency, and first responders must know how and when to work with each other. Second, there must be a plan of action that has been developed and agreed to by all participants. This must happen long before an emergency develops. The plan must be practiced and refined as needed. All members of the community safe drinking water team must be actively engaged; everyone becomes responsible for the final outcome.

TEAM BUILDING

Roles of Water Utility, Public Health Agencies, and First Responders

The process begins by requesting formal cooperation and encouragement from your organization's senior management. The next step is to formulate a list of stakeholders that must be involved. Start with high-ranking individuals; they can, if necessary, delegate someone from their staff. The key to developing this list is being familiar with and understanding the agency and nonagency boundaries and functions, as well as regulatory jurisdiction, for your particular community. In most areas, there are multiple agencies or entities that supply and treat water and have responsibility for some aspect of public health protection. Keep the list concise, identifying people with authority as well as knowledge. The following list is an example of those who should be included on the core team:

- Water utility:
  - Chief engineer
  - Director of operations
  - Director of communications
- Public health agency:
  - Medical officer of health
  - Director of communicable diseases
  - Public health inspectors
  - Director of public health laboratory
  - Social workers
  - Director of communications
- First responders:
  - Police chief*
  - Fire chief
  - Director of emergency medical services

---

* Because of heightened security around the world, the police chief should also serve as a liaison among all local and regional law enforcement agencies involved should the situation be criminal or terrorist related.

- Environmental regulators (provincial/state level; federal, if necessary):
  - Approvals engineer
  - Compliance inspector and/or investigator
  - Director of analytical laboratory
- Elected officials (local, provincial/state, federal)

This will be your core safe water team. As you begin to work as a team, other members may be identified who have a particular expertise and should be invited to join.

A word of encouragement is necessary at this point. It can be challenging and frustrating to establish working relationships and common goals with external agencies or entities that you do not usually work with. Throughout the development of a team approach, team members must be reminded often that there is only one goal: the health and well being of its customers/citizens. The ultimate objective is still safe drinking water for the entire community. Although this seems obvious, there may be times when members of the safe water team should be reminded to repeat this mantra. All personal or agency-specific issues must be put aside whenever the health of individuals or the community is at risk.

The next step is to contact the appropriate members of your team and set up a meeting time and location. You should prepare an initial agenda that will serve as your guidepost for the meeting. An agenda format follows.

Agenda (initial)

a. Introductions: Give each person an opportunity to introduce themselves and to outline their involvement on the team.
b. Notification list: Provide a format for each team member to list appropriate contact information during business and nonbusiness hours.
c. Alternate members: Identify an alternate for each team member and include their contact information on the notification list.
d. Chair: Elect one person to chair the meeting and distribute housekeeping tasks on a rotating basis (e.g., minute taking and meeting preparation).
e. Laws and regulations: Distribute appropriate public health and water regulations.
f. Mission statement: Develop a brief statement that captures why the team members believe they are there and what they expect to accomplish.
g. Agency responsibilities: Initiate a dialogue among members to discuss who does what, when, and where. This is the first step in organizing a crisis management plan.
h. Future meetings: Set up regular meetings, such as every month, and vary the locations so that members become familiar with each other's place of business. NOTE: "Brown bag" lunch meetings can mitigate busy schedules.

COMMUNICATIONS MANAGEMENT

Pre-event Planning (Normal Situation)

After accomplishing the first meeting and setting the team in motion, the team should focus on two issues:

1. Developing a crisis management plan

2. Creating and sharing a common language and knowledge base, which will assist team members in routine issues as well as during an actual crisis

Crisis Management Plan

Below is a suggested to-do list that will assist your team in preparing and drafting a crisis management plan. Once the plan is in effect, it is extremely important that the document be reviewed and updated yearly, and that annual desktop exercises be organized and conducted.

- Designate a name for your team such as "emergency communications team (ECT)."
- Distribute and review existing plans from other agencies, companies, etc.
- Establish an outline or sections for the plan.
- Assign sections to be drafted to members.
- Designate a crisis event spokesperson. This should be someone from the public health agency, because the public will see this as a health issue rather than solely a water utility infrastructure issue. A support panel made up of designated individuals from key stakeholder agencies (e.g., water utility, public health and environmental regulators, and emergency responders) should also be present to answer technical questions, if necessary. Discussion prior to the public statement release should be a joint effort among these key agencies.
- Set up "watchdog criteria" to track key events from a heightened alert to an emergency situation. Include information such as number of people impacted, area of impact, assessment of potential severity to public health, suspicious origins, length of time before situation is corrected, etc.
- Develop a set of guidelines or protocols for the ECT involving communications to elected officials, media, other agencies, and the general public.
- Identify "precrisis" warnings that usually predict a problem in the system and the triggering factors that raise the situation from normal, to heightened, to emergency levels.
- Locate a crisis center that will serve as a meeting place for the ECT. This center should have all pertinent office equipment and communication tools.
- Develop draft press releases, fact sheets boil-water advisories, no-water advisories, public health guidelines on waterborne illnesses, etc. These drafts should include translations in appropriate languages.
- Maintain a notification list for comprehensive internal agency outreach, such as switchboard operators, public relations staff, and other employees that may be called upon for assistance.
- Outline procedures for requiring a boil-water advisory and when that advisory should be rescinded. Do the same for no-water advisory.
- Develop specific outreach methods for special risk groups, such as people with special needs, sensitive communities, food establishments, etc.
- Explore toll-free hotlines and a "rumor control" desk.
- Identify possible relocation facilities and alternate drinking water sources.
- Create a quick reference guide showing the order of steps to follow during an emergency (see Table 4.1).

Table 4.1
Key steps in effective crisis management

1.	Establish the crisis center location.
2.	Set a meeting time and notify the emergency task force members.
3.	Prepare an agenda and a brief description of the situation.
4.	Establish goals and objectives for managing the crisis and what can be done to mitigate the crisis.
5.	Employ protocols for contacting outside agencies and appointed or elected officials—implement agency contact call-down list.
6.	Identify a command center team, who will designate a spokesperson.
7.	Review and finalize the draft key messages (e.g., description of event, population and municipal area affected, required public health precautions, updates), fact sheets, press releases, frequently asked questions, etc.
8.	Issue appropriate public announcements, press releases, and advertisements (e.g., billboards), and update Web site, etc.
9.	Instruct telephone operators, receptionists, safety officers, and others, as to how to respond to the public.
10.	Upgrade telephone system, if necessary, and install a toll-free hotline.
11.	Develop a system for rotating key staff.
12.	Update employees on a regular basis.
13.	Set up and monitor a rumor control desk.
14.	Identify one spokesperson to interact with the media and set up an in-house public relations team.
15.	Set up a series of scheduled meetings at the crisis center among the various agencies or the public, as necessary.
16.	Establish a pressroom location and arrange a press conference.
17.	If a boil-water advisory is in effect, issue notices and fact sheets. The public health agency should take the lead on health-related public statements.
18.	Notify special needs and sensitive communities.
19.	Monitor radio and TV stations for information on the situation.
20.	In the final stages of the crisis, determine whether a follow-up press release is necessary and establish criteria for determining the end of the crisis.

Common Language and Knowledge Base

Communications is a fundamental issue for the ECT. The team must learn how to communicate with and understand each other, especially during an emergency. In addition to the regular meetings already established, the following list of ideas can help instill trust, understanding, and loyalty within the team. Team members should set up a regular program that would include the following elements:

- Develop a glossary of terms that each agency uses on a daily basis.
- Discuss your area's geographical regions and how the local water distribution infrastructure supplies those areas.
- Observe maintenance and/or proactive procedures in the field, such as a flushing program.
- Set up a tour of the water-testing laboratory, water source, pump station, etc.
- Discuss water and health issues that the public is concerned with, such as risks associated with disinfection, disinfection by-products and cancer, or "how clean is clean?"
- Reach out to agencies in surrounding areas and share common experiences and lessons learned.
- Look at ways to be proactive, such as notifying schools to flush their indoor plumbing just before the school year begins.
- Visit infrastructure emergency sites such as a water main break location.
- Develop joint educational materials on water distribution systems, water treatment, and health-related issues.
- Make joint presentations to schools, community groups, churches, environmental groups, etc., to discuss water- and health-related issues. (A side benefit is that the community as a whole gets to know its water and public health agency individuals).
- Participate jointly in related training and conferences that are offered to water and public health agencies.
- Write a paper for joint publication or presentation.

Heightened Alert Situation

Once a heightened alert situation occurs, such as an unusually high incidence of community enteric disease, a significant infrastructure failure, or a water treatment failure, the ECT contact person is notified and, in turn, immediately notifies the other ECT members. It is important to have current 24/7 contact information for all core ECT members. This initial contact can be done effectively via a telephone conference call. It is important that all members hear the same message and understand the situation. A report of the situation should be given by the appropriate agency. Updates should be reported hourly so that the ECT can monitor the impacts and determine whether the alert level should be raised to an emergency status. In the meantime, the ECT will perform the following tasks:

- Advise their appropriate superiors.
- Implement the watchdog criteria.
- Set up a meeting time at the command center.

- Begin review of fact sheets, press releases, media contacts, etc.
- Clear any prescheduled, nonemergency events from all individual calendars.
- Review the crisis management plan.
- Identify support staff and resources such as computers, cell phones, pagers, etc.
- Notify relocation experts and real estate agents if a large population will be impacted.

An important issue is determining whether, and when, public notice is necessary. There are no hard and fast rules; the decision will involve a collaborative effort among the ECT members. The public's right to know must be balanced carefully with the facts that surround each situation. Establishing trust and understanding among its members will allow the ECT to make a decision as a unit that is based on facts. To assist the team, the watchdog criteria should be reviewed and followed. However, in no case should the health and well-being of the community be compromised.

Emergency Situation

During a heightened alert situation, the ECT meets at the command center, where one of two situations will develop:

1. The heightened alert situation is brought under control or contained; thus, there are only limited or isolated impacts to the population. Public notification is targeted and initiated immediately.
2. The heightened alert situation is now an emergency and has widespread health impacts within the municipality. The full crisis management plan is triggered.

The ECT will initiate the full crisis management plan from the command center. A secondary site should be established as a media center and should be offsite from the actual crisis location. This is where the media will go for more information. The spokesperson at the media center must be given very specific instructions and information. Updates will be communicated to the media and others only from the media center; the command center must retain its unique status as the central location where all action is to be determined and information passed on to the crisis location and media center. This is an important distinction, as you do not want different messages being delivered, that could cause harm to the overall message sent to the media and the public.

During an emergency situation, a "feeding frenzy" for information and updates often develops among the media. All public health agency personnel must remain calm. The designated spokesperson, usually from the public health agency, should deal directly with the media, ensuring one clear and accurate message to the public.

When the situation returns to normal status, the ECT should meet for an incident debriefing so that strengths and weaknesses can be discussed, evaluated, and documented. The results of the debriefing may lead to revision of the crisis management plan. This process will help you and your agency be better prepared for any future emergencies.

CHAPTER 5
PARTNERSHIPS TO MANAGE RISKS FROM INFRASTRUCTURE FAILURES

Karen M.E. Emde, Daniel W. Smith, James A. Talbot, Les Gammie, and Nelson Fok

INTRODUCTION

The cornerstone for any collaborative program to prevent and/or manage health risks that could be caused by infrastructure failure starts by first sharing knowledge of the occurrence (e.g., type of failure, magnitude of failure, precise geographic location, dates and/or times of failure events, and/or resulting exposure in the community). Without those first pieces of critical information, it can become difficult or nearly impossible to conduct effective risk management and community surveillance to perform the following tasks:

- Assess and rank contamination potential, and determine the contaminant identity, quantity, movement, and/or amplification through the municipal distribution system (e.g., entry due to backsiphonage of hazardous or infectious agents from industries and/or other sites directly connected and in proximity to the infrastructure failure site, and infiltration of soil-borne contaminants in close proximity).
- Conduct surveillance to detect, quantify, and characterize suspected or confirmed waterborne contaminants.
- Determine the extent and patterns of community exposure, and assess, rank, and link host symptoms of illness with suspected waterborne contaminants and the infrastructure failure event.
- Assess, evaluate, and link the vehicle of waterborne contaminant exposure, including but not limited to direct exposure to contaminated water versus secondary exposure from other possibilities such as contaminated food and beverages, recreational facilities, person-to-person transmission, oral–fecal transmission, zoonotic transmission, etc.
- Implement short- and/or long-term incident response procedures, such as boil-water advisory, no-water advisory, provision of alternate sources of water, mains repair, disinfection, increased sampling, and flushing efforts, etc.
- Implement appropriate incident and/or emergency plans.
- Implement short- and/or long-term health surveillance monitoring to detect, treat, and track exposed individuals.
- Inform medical providers of the potential for increased cases or casualties due to the infrastructure failure, as well as the nature and source of the suspected or confirmed waterborne contaminant to aid with diagnosis and treatment.
- Inform first-response agencies of physical/biological/chemical hazards caused by, or as a direct consequence of, the infrastructure failure event, as well as the potential for community risk and/or harm.
- Inform local, provincial/state, and/or federal stakeholder agencies, politicians, affected communities including special risk groups (e.g., hospitals, dialysis patients, and food-processing plants), and the media of the event and its significance, as well as any necessary precautionary measures.

PARTNERSHIPS TO MANAGE RISKS

Determining the exposure source and gathering critical supporting evidence for waterborne contaminant exposure and symptoms of community illness, irrespective of cause of failure, requires vigilance, persistence, and, more often than not, the sleuthing experience, intuition, and the occasional lucky break of a master detective.

Whether the outbreak is only suspected or actually confirmed, water utilities and public health agencies immediately implement their short-term investigation and correction procedures, which are designed to verify, contain, manage, and ultimately eliminate the threat to the infrastructure as well as the community. Figure 5.1 illustrates some of the steps involved in short-term incident response and risk management for situations when waterborne contaminants enter the municipal water system because of a failure event.

Long-term corrective measures to prevent future outbreaks typically occur as a result of retrospective analysis and evaluation of the following issues:

- Circumstances and causes of infrastructure failure
- Appropriateness of response and effectiveness of corrective action taken
- Consequences of the failure in light of the costs of community short-term and long-term waterborne disease effects (e.g., costs of medical treatment, loss of wages, loss of productivity), as well as community economic impacts that include but are not limited to loss of business and product for industries using municipal water as part of their product or a service such as food production, light and heavy manufacturing, the hospitality and tourism industry, recreational facilities, sports facilities, educational institutions, medical institutions, business organizations, etc.
- Costs of investigation and remedial measures to control the incident or emergency caused by the failure event
- Costs of short-term and long-term loss of business and reputation for the water utility, as well as potential costs of liability

Long-term corrective measures can involve changes in the following areas:

- Operations (e.g., changes to specifications for infrastructure materials and vendors, design and construction, and maintenance and renewal procedures)
- Analytical methods to improve the detection and quantification of suspected waterborne contaminants or development of new field and laboratory methods
- Regulatory standards and expectations
- Response procedures
- Utility and/or public health surveillance programs, etc.

Figure 5.2 illustrates steps in the cycle of long-term response strategies to waterborne disease outbreaks.

The ultimate target, or gold standard, would be to develop, implement, and maintain a joint risk management program whereby each member agency/organization retains organizational

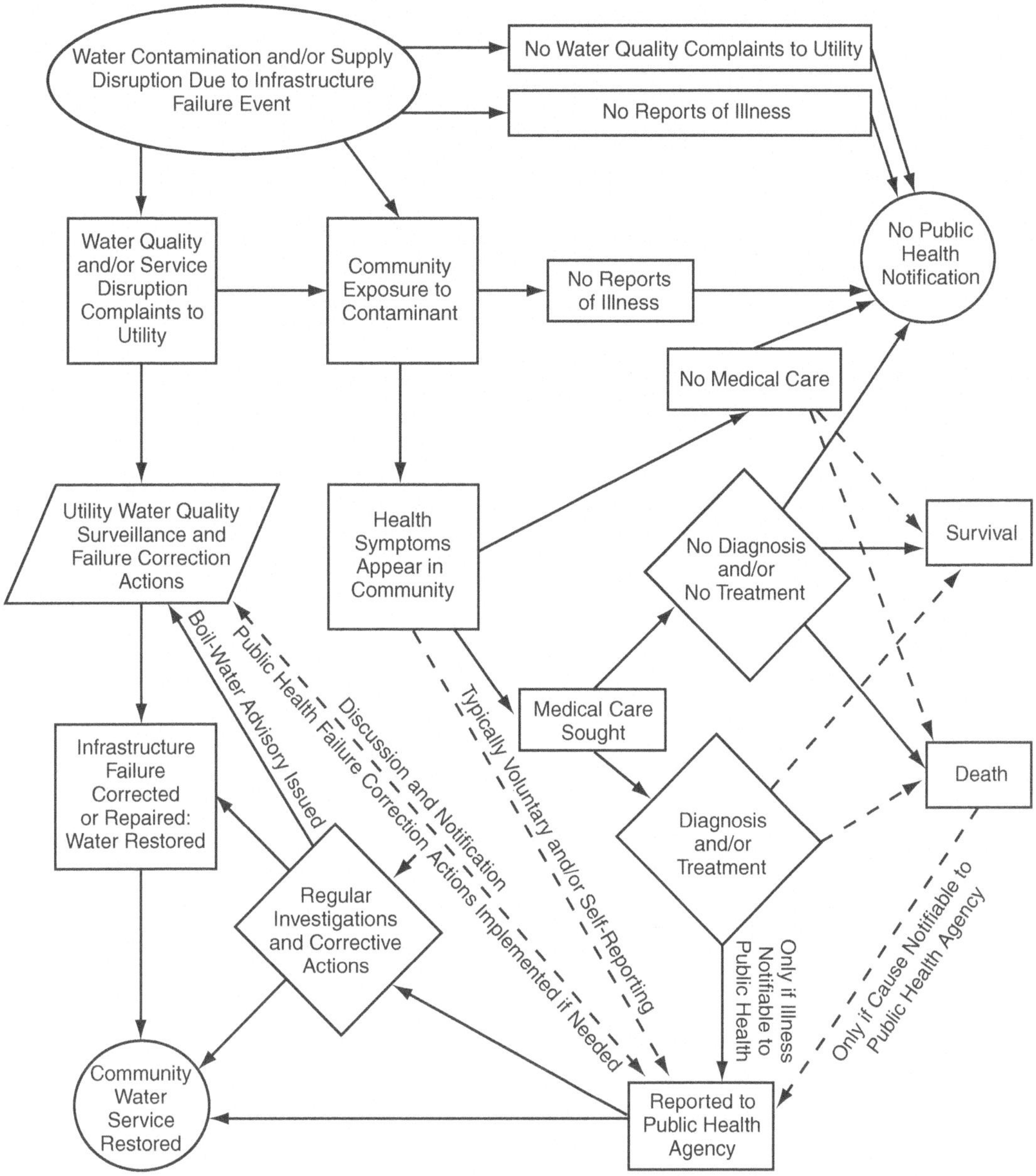

Figure 5.1 Immediate response to manage health risks due to infrastructure failure

and jurisdictional autonomy, but there is consensus, commitment, and willingness to operate in an open, collaborative, and active fashion with other critical community stakeholder agencies. Figure 5.3 illustrates some of the key activities that are involved in a joint risk management partnership. Chapters 2, 3, and 4 in this report suggest activities that can be used to initiate a joint risk management program, as well as strategies to sustain the program and level of collaboration. This open relationship requires timely sharing of critical information so as to contain and manage the situation and to prevent future occurrences.

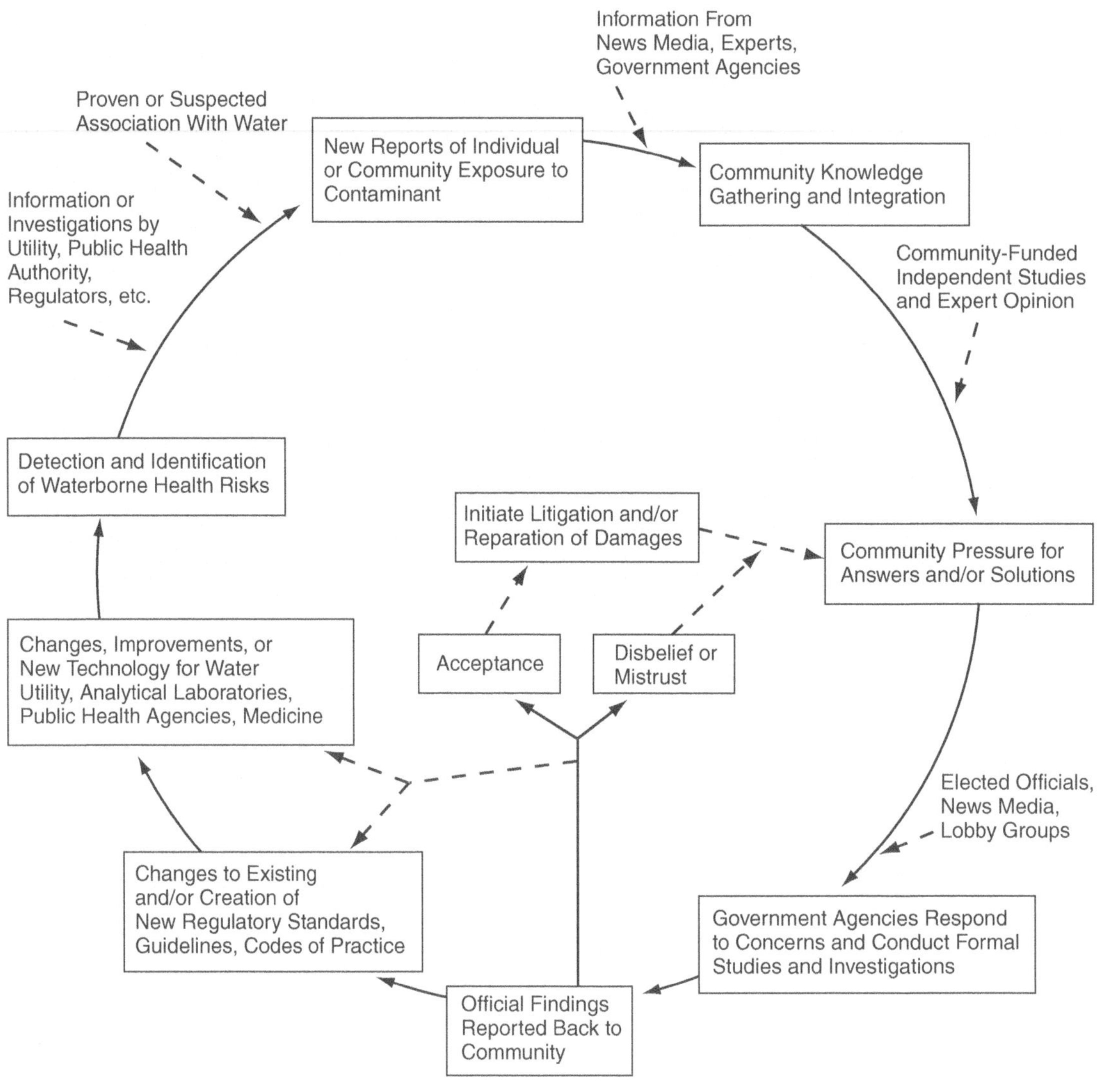

Source: Adapted from Emde et al. 2001, Brown 1992.

Figure 5.2 Cycle of long-term response to waterborne disease outbreaks

ANATOMY OF WATERBORNE DISEASE OUTBREAKS: COULD PARTNERSHIPS HAVE HELPED?

Two case studies of well-characterized waterborne disease outbreaks were evaluated, based on available historical evidence, to determine whether a shared risk management approach could have led to earlier detection and, thus, better management or even prevention of the outbreak. The first case study, the 1993 Gideon, Mo., *Salmonella typhimurium* outbreak, is typical of waterborne disease outbreaks caused by insidious, long-term, undetected pathogen intrusion into the municipal water due to loss of structural integrity of key infrastructure components (Clark et al. 1996a, b).

The second case study, the 1964–1979 Woburn, Mass., cancer cluster, illustrates consequences of low-level, continual infrastructure failures and contaminant intrusion, and poor judgment

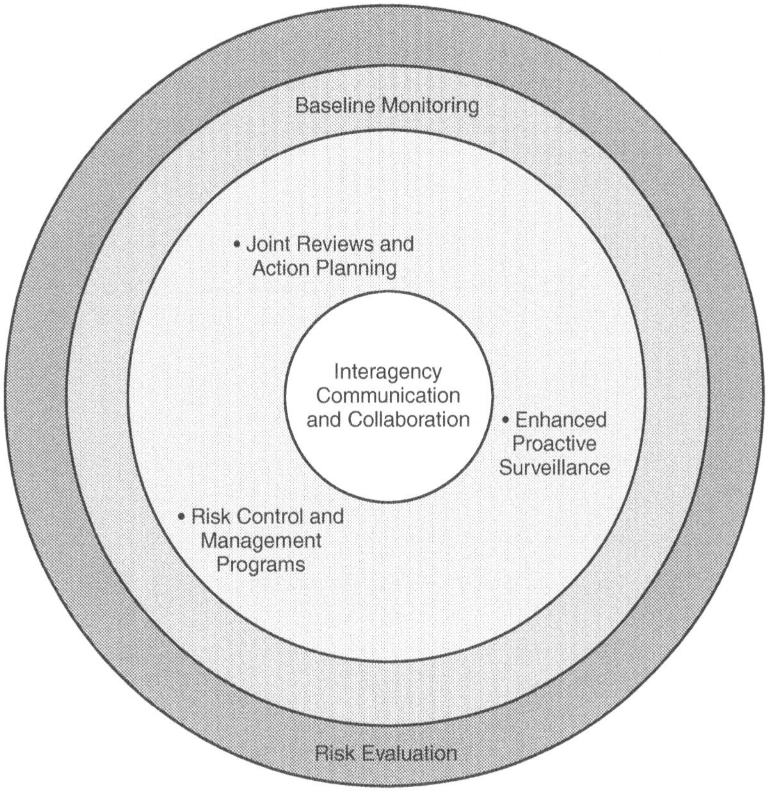

Figure 5.3 Some key activities involved in joint risk management

in siting and monitoring of municipal water wells, which resulted in intermittent waterborne exposure to various carcinogens during the 15-year time span that the suspect wells were in service (Rowe 1981, Murphy 1990).

Both cases provide illustrative examples of where improved, effective communications and/or collaboration between key community partner agencies could have improved the final outcome; that is, fewer cases of waterborne illness, fewer infrastructure failures, and, ultimately, safer drinking water.

Although both outbreaks occurred in smaller communities and water utilities, the pattern of events leading up to the outbreak/chronic disease cluster and actions taken are not exclusive to small centers. Outbreaks in large centers and/or water utilities have been shown to follow similar patterns.

Waterborne Infectious Disease Outbreak Caused by Infrastructure Failure: 1993 Gideon, Mo. (USA), *S. typhimurium* Outbreak

Although there are innumerable published reports of outbreaks caused by waterborne, infectious organisms, few have been as rigorously analyzed and had findings published as the 1993 Gideon, Mo., outbreak. This analysis of the outbreak will examine published records, early warning signs, what was and/or was not done by the agencies involved, and, finally, from a retrospective standpoint, what actions or partnerships could have minimized the severity and impact of this outbreak.

Details from the Gideon outbreak and a timeline of events are shown in Table 5.1.

Table 5.1
Details and timeline of Gideon, Mo., waterborne infectious disease outbreak

	Water Profile
Municipal water supply and distribution system (DS):	Contains two deep (approximately 1,299 ft [396 m]) municipal wells; unchlorinated; groundwater supply overlies a geologically active thermal fault, thus causing higher groundwater temperature (approximately 57°F [14°C]). During high municipal flow conditions, pressure in wells drops significantly. Wells are operated in an alternate fashion on a monthly basis. The municipal DS was constructed in the mid-1930s and consisted of small-diameter (e.g., 2, 4, or 6 in. [50, 100, or 150 mm]) unlined steel and cast-iron mains. The town has two municipal water storage towers (50,000 and 100,000 gal [0.19 and 0.38 ML] capacity). There was also a private industrial tank connected to the municipal system.
	Public Health Initial Involvement
Outbreak first noticed by public health department (PHD):	Nov. 29, 1993—Two local high school students hospitalized with culture-confirmed salmonellosis. Nov. 29, 1993, to Dec. 1, 1993—Five additional residents (1 student, 1 day-care client, 2 nursing home residents, 1 nursing home visitor) also hospitalized with salmonellosis. Nov. 29, 1993, to Jan. 8, 1994—Thirty one laboratory-confirmed cases of salmonellosis; 7 patients died.
Total cases reported to PHD:	November 1993 to Dec. 27, 1993—Approximately 600 cases diarrheal illness (town population approximately 1,104). • Crude attack rate: 54% • Crude mortality rate: 0.63%
Confirmed etiologic agent:	*Salmonella typhimurium*
	Water Utility Initial Involvement (Independent of Initial PHD Involvement)
Early warning signs:	November 1993—Increased water taste and odor (T&O) complaints after a routine mains flushing program involving all 50 hydrants in the municipal system. DS sampling to detect causes of T&O showed elevated HPCs as well as an increase in total and fecal coliform-positive samples.
Utility initial responses to correct T&O complaints:	November 1993—Intensified sequential mains flushing efforts. Increased DS sampling.
	Involvement by Other Agencies
Initial PHD investigation:	November 1993—Missouri Department of Health (DOH) investigated initial salmonellosis cases and determined that the likely cause of the outbreak was consumption of municipal drinking water. The Missouri State Public Health Laboratory identified initial isolates as dulcitol-negative *Salmonella.* The Centers for Disease Control and Prevention (CDC) identified initial case isolates as *S. typhimurium.*
Initial PHD notification to other agencies:	December 1993—The DOH informed the Missouri Department of Natural Resources (DNR) that the Gideon outbreak was thought to be waterborne. Dec. 17, 1993—The DOH informed the CDC that the Gideon outbreak was due to contaminated municipal water.

(continued)

Table 5.1 (Continued)

	Dec. 22, 1993—The DOH invited the CDC to join the investigation. Jan. 8, 1994—The DOH identified 31 laboratory-confirmed cases of salmonellosis during the time period from November 1993 to Jan. 8, 1994.
DNR investigation and actions taken:	Dec. 17–21, 1993—The DNR begins additional water testing and finds total and fecal coliform-positive water samples as well as elevated HPCs. Dec. 18, 1993—The DNR issues a boil-water advisory. Dec. 23, 1993—The DNR places an online chlorinator on the municipal well. Dec. 30, 1993—A privately owned, industrial water tank was permanently disconnected from the municipal supply.
Joint DOH and DNR activities:	Dec. 29, 1993—The DOH and DNR jointly produced a flyer explaining the boil-water order, which was placed in mailboxes of all homes on Dec. 29, 1993.
Involvement of federal agencies:	Jan. 4, 1994—The USEPA field team, working with the CDC and State of Missouri, began field investigation (i.e., sanitary survey of Gideon municipal water system, sample collection, and analysis). Samples were taken throughout the DS (building and/or hydrant samples), including private and municipal water tanks and sediments.
Sanitary survey findings:	Well site integrity: • Agricultural runoff was present from adjacent fields in the well-shed. • The well pits routinely became flooded with runoff and had to be pumped before the water meter could be read. • There were no apparent backflow devices on well pumps. DS integrity: • There was significant tuberculation and corrosion in mains. • There were inconsistencies in free chlorine readings throughout the DS. Storage tank integrity: Municipal and private tanks had holes in roofs; there was evidence of corrosion and minimum maintenance; the tank roofs served as a local pigeon roost; bird droppings, feathers, etc., were observed in the municipal water that was stored inside the tanks, and outside the tanks as well.
Key Evidence Confirming Gideon Municipal Water as the Cause of the Outbreak	
Water sample evidence:	Municipal water samples taken at sites near a school and nursing home tested positive for *S. typhimurium* during the second flushing program. *S. typhimurium* was isolated from the riser pipe of a private water tank. Other distribution samples showed elevated HPCs, positive total and/or fecal coliforms, and reduced chlorine residual.
Hydraulic model evidence:	Retrospective hydraulic modeling traced the origin and then the movement of water from the contaminated municipal tanks through the DS.
Other possible sources:	The private water tank was ruled out when follow-up pipe rehabilitation determined that the backflow prevention valve was functional.
Additional contributing factors:	The Gideon water utility was found to lack proper resources (e.g., financial, physical) necessary for proper infrastructure repair and maintenance.

Source: Adapted from Geldreich 1996; Clark et al. 1996a, b; Clark and Grayman 1998.

Waterborne Chronic Disease Cluster Linked to Infrastructure Failure: 1964 to 1979* Woburn, Mass. (USA), Cancer and Childhood Leukemia Cluster

The story of the Woburn, Mass., childhood leukemia cluster is best known as the subject of the book by Jonathan Harr (1995) and major motion picture by Robert Redford and Disney Studios, both titled *A Civil Action* (Zaillian 1999). Between the years 1964 to approximately 1987 there were 28 cases of childhood leukemia identified in Woburn; 16 children died (Latowsky 1998, BEHA 1997). Additional studies identified elevated rates of adult kidney cancer; male breast cancer; heart, immune, and nervous system disorders; and birth defects in Woburn during that same time period.

Two municipal groundwater wells, G and H, were determined to be grossly contaminated with a variety of chlorinated organic chemicals, pesticides, and metals, and were finally closed in 1979. This case, together with the Love Canal (City of Niagara Falls, N.Y.) case, helped establish the USEPA Superfund program in the early 1980s (Murphy 1990).

Direct exposure to drinking water from wells G and H was determined to be the cause of the childhood leukemia cluster in Woburn. Epidemiological and risk exposure studies of the site continue today. The site of wells G and H as well as the local industrial park (Industriplex) were later declared national priority sites under the Superfund program (Murphy 1990).

Tracking chronic disease caused by drinking water contamination and infrastructure failure is extremely complex, much more so than tracking waterborne infectious disease. Often, there are multiple sources of environmental and occupational exposure. These exposures may arise from industrial, municipal, agricultural, or personal sources (e.g., home-building materials, home use of pesticides and herbicides). The latency period for symptoms of chronic disease is typically long, on the order of a few years to tens of years. If a chronic disease cluster or outbreak is linked to a waterborne exposure, tracking exposed individuals can be equally complex, given the long latency period and relative mobility of individuals and families over the term of that latency phase.

Epidemiological findings of the Woburn childhood leukemia cluster determined that the affected children were exposed in utero to an intrauterine dose of waterborne carcinogens, as well as postnatal exposure from ingestion and/or dermal exposure to water from wells G and H (Rowe 1981). Though childhood cancers are still somewhat rare, it has been determined that rates of childhood cancers in the United States have increased almost 21% during the time period from 1975 to 1998 with *in utero* and early childhood exposure to environmental substances playing a significant role in development of disease (Gouveia-Vigeant and Tickner 2003). The recent USEPA draft final *Guidelines for Carcinogen Risk Assessment* have determined that children younger than 2 years old are 10 times more susceptible to chemical carcinogenesis than adults, whereas children from 2 to 15 years old are 3 times more susceptible than adults (USEPA 2003).

Particular details and a timeline of events in the Woburn case are shown in Table 5.2.

---

* Although the wells were disconnected in 1979, childhood leukemia cases were reported until 1987 because of the lag period between exposure and onset of attributable clinical symptoms.

Table 5.2
Details and timeline of Woburn, Mass., waterborne chronic disease cluster

	Drinking Water Characteristics	
Municipal water supply:	The supply consisted of a series of shallow wells: six wells (A to F) were located near Horn Pond, and two wells (G and H) were constructed along the east flood plain of the Aberjona River.	
	Woburn had a history of problems with water quality and supply during the 20th century and had to change location of wells, pumping equipment, and supply several times (Tarr 1987).	
	Woburn wells A to F also had numerous water quality complaints, as did Horn Pond, during the same period as wells G and H, although wells A to F were supplied by a different aquifer than wells G and H (Tarr 1987).	
Woburn DS:	The system consisted of a series of eight municipal wells and two significant booster pumps that delivered water through a pipe network to consumers (system used pump control valves, water storage tanks, and reservoirs to regulate distribution).	
	Flow from wells G and H was significantly affected by pumping of well E (Murphy 1990).	
	Pipes were primarily cast iron with some asbestos–cement, and ranged in diameter from 6 to 24 in. (150 to 600 mm). They were installed between 1900 and 1977 (Murphy 1990, Tarr 1987).	
	History of Suspect Wells	
	Well G	Well H
Constructed:	1964	1967
Depth and diameter:	88 ft (26.8 m) deep; 24 in. (600 mm) borehole	89 ft (27.1 m) deep; 24 in. (600 mm) borehole
Backflow valves to prevent siphonage when one well is turned off and the other is pumping:	Not apparent	Not apparent
Groundwater aquifer:	Aberjona	Aberjona
Well production:	~800 gpm (~50.5 L/sec)	~700 gpm (~44.2 L/sec)
Chlorination system:	Installed in 1968	Installed in 1968
Well use:	Intermittent	Intermittent
Primary water supply for:	East Woburn (when wells were in use)	
Reasons for reinstating wells:	• Hot, dry summers in the early 1970s caused significant drops in production by wells A to F, restricting the available municipal supply. • Increased residential and industrial development in the 1970s created additional water demand that wells A to F could not supply.	
Water quality complaints:	• Constant T&O, color, and hardness complaints when wells G and H were in use. • Red water complaints also.	
Wells closed:	1979	1979
Estimated total days of well production and potential exposure to water:	~2,995 days during a 15-year period	

(continued)

Table 5.2 (Continued)

Water Pollutants Found in Wells G and H in 1979 (Chronic Diseases Division 1981)

	Well G		Well H	
Sampling date	May 14, 1979	Sept. 24, 1979	May 14, 1979	Sept. 24, 1979
Chloroform (ppb)	11.8	1.1	ND*	ND
Trichloroethylene (ppb)	267.4	117.6	183.6	63.0
Tetrachloroethylene (ppb)	20.8	18.3	13.4	9.0
1,1,1-Tetrachloroethane (ppb)	0.6	ND	ND	2.1
Dibromochloromethane (ppb)	2.0	ND	ND	ND

Cancer Cluster Details

Suspected environmental exposure period:	1964 to 1979—Approximately 2,995 days, varying from intermittent to constant exposure, during a 15-year period)
Details of childhood leukemia cluster:	• Seventeen cases of acute lymphocytic leukemia were diagnosed during the exposure period. • Three cases of acute myelocytic leukemia were diagnosed during the exposure period. • One case of chronic myelocytic leukemia was diagnosed during the exposure period. • Between 1982 and 1990, seven additional childhood leukemia cases were diagnosed (five male, two female). • No additional new cases were diagnosed between 1987 and 1992.
Details of suspected increase in Woburn adult cancer during exposure period:	Various epidemiological studies have been conducted from 1979 to the present. Early studies found statistically significant increases in rates of kidney and female reproductive organ cancers (other than uterine cancer), and higher rates were noted for adult leukemia, and for prostate, breast, lung, pancreatic, and stomach cancers. All studies acknowledge the limitations in interpretation due to the relatively small number of cases involved (Rowe 1981, Kotelchuck and Parker 1979).

Significant History About the Site of Wells G and H

Municipal activity:	1958—An engineering consultant hired by the City of Woburn cautions that the Aberjona River Valley is too polluted to be considered for a public drinking water supply. 1968—Chlorination facilities were installed first at well H, then at well G. 1969—The city announces plans for construction of a new industrial park (known as the Industriplex) immediately north of wells G and H. 1974—An extensive study of Horn Pond (the approximate site of wells A to F) revealed significant pollution, including elevated levels of chlorides, nitrates, phosphates, coliforms; low dissolved oxygen; and excessive growth of aquatic weeds and algae. 1976—An expert committee was established to study the water problems at wells G and H (specifically, odor and color). A consultant hired by the City of Woburn maintained that water quality problems at wells G and H were due to interaction between chlorine added to the wells and manganese in the DS. The consultant recommended air injection in perimeter wells to correct the problem (Tarr 1987).

(continued)

Table 5.2 (Continued)

Massachusetts DOH activity:	1967—DOH recommends that wells G and H be removed from service because of poor bacterial quality of the water.
	For wells to remain in service, DOH requires continuous chlorination of wells G and H.
	1968–1969—DOH notes that the proposed site for the Industriplex in the Aberjona River flood plain and immediately north of wells G and H is significantly contaminated.
	DOH notes that the Industriplex site was formerly occupied by the Stauffer Chemical Co.
	DOH notes that filling and draining of the site to accommodate construction needs could have adverse effects on the town's water supply, especially during high, or peak, river flow.
	DOH recommends that abandoned waste lagoons and dumps of Stauffer Chemical Co. be excavated and disposed of before construction of the Industriplex complex.
	DOH requests that plans for site sewage disposal and a lift station be first approved by DOH before construction.
	Early 1970s—DOH requests Woburn to not use wells G and H because of poor water quality (e.g., elevated levels of nitrates, ammonia, chlorides, sulfates, sodium, manganese, hardness, color, odor, turbidity, and sediment).
	Analytical technology for various chemical contaminants was continually defined and refined for use in drinking water. The 1976 edition of *Standard Methods for the Examination of Water and Wastewater* (APHA, AWWA, and WPCF 1976) first mentions a proposed method, an organics–carbon adsorbable method, for the identification of organic contaminants but at that time was not specific to chlorinated solvents in drinking water.
	Late 1970s—Nationwide publicity appears about the Love Canal, N.Y., toxic waste dump and its link to contaminated groundwater and childhood cancers, as well as initial reports of increased childhood cancers in Woburn.
	May 1979—Wells G and H in Woburn are permanently closed.
Post–May 1979 events:	In Woburn, further USEPA investigations and epidemiological studies have continued at the wells G and H and the Industriplex sites. Extensive sampling of soil, sediments, groundwater, and surface water has been conducted using the best available detection and measurement technology.
	These USEPA studies found diverse mixtures of volatile organics, semivolatile organics, and inorganic metals in significant concentrations in all types of samples taken from the investigation site.
	1982—USEPA established a Superfund site remediation program, primarily because of the revelation of the situations in Love Canal, N.Y., and Woburn, Mass. The sites of wells G and H and the Industriplex site were identified as one of the largest hazardous waste sites in the United States and therefore designated as Superfund sites.
	The Superfund program spurred growth and provided significant funds research into the extent of hazardous contamination from former industrial sites, which resulted in improved methods of detection and quantification, and a better understanding of the routes and mechanisms of entry into the body as well as in short- and long-term *in vivo* effects.
	Eight Woburn families filed lawsuits against three local industries thought to be responsible for environmental contamination and the increase in childhood leukemia morbidity and mortality in Woburn (*Anne Anderson et al. v. Cryovac et al.*[†] 1983).

* ND = not determined.

† Cryovac Inc. was a division of WR Grace & Co.

POSTMORTEM ANALYSIS AND RETROSPECTIVE REFLECTION

What do the outbreaks in Woburn, Mass., and Gideon, Mo., have in common with each other, as well as with other known waterborne disease outbreaks caused by infrastructure failure or other types of water utility failure? The first similarity for these two waterborne disease outbreaks, as with so many others, was failed communications, that is, failure to communicate within organizations as well as among organizations with overlapping responsibility (i.e., issues of jurisdiction and efforts at protection of that jurisdiction).

Had the consultant hired by the Woburn water utility in 1958 to advise on the suitability of sites for new and urgently needed water wells been heeded, perhaps the Woburn outbreak may never have occurred. At the very least, the pattern of community morbidity and mortality caused by this outbreak would not have been as a result of consumption of drinking water from wells G and H. If there had been better communication and respect for the opinion of the Massachusetts DOH in the 1960s when warnings were issued about Woburn's water quality and the possible outcomes, perhaps the tragedy of the late 1970s could have been avoided. If the planning of future subdivisions and municipal expansion had included the foresight of predicting the availability of adequate, safe drinking water rather than development for purely economic considerations, perhaps wells G and H may have remained closed. Had the detection and analytical technology that was developed in the 1970s been available for practical, routine use by water utilities, then the effects of chlorination of organic, waterborne contaminants may have been detected earlier and remedial measures could have been implemented. Certainly, had the DOH realized that the proposed remedy for handling unacceptable coliform counts of wells G and H—increased chlorination—created a greater health risk, alternate remedial measures could have been explored. This technology was not routine in the late 1960s. Hindsight can be perfect.

Similarly in Gideon, Mo., could the outbreak have been prevented or mitigated by early notification to the local public health department of the elevated levels of coliforms and the increase in localized customer complaints? Could the distribution system flushing program that was implemented for increased T&O complaints and total coliform occurrences have been better executed to minimize community exposure? Would earlier issuance of a boil-water advisory, in response to the significant increase in coliform-positive samples as well as other measured water quality parameters, have reduced and/or prevented morbidity and mortality? Significant increases or decreases are defined with respect to the historical and/or measured baselines for a community.

So many questions, yet these remain the very same questions and issues facing water utilities and public health agencies today. The gold standard for drinking water quality is the total absence of waterborne disease (Eliassen and Cummings 1948, Grabow 1996). Both outbreaks presented in this chapter provide clear evidence that morbidity and mortality remain compelling and conclusive measures of water safety but are not the desired indicators of choice for routine measurements to prove that drinking water is safe.

Mortality and morbidity from waterborne disease can be prevented by good multibarrier practices, including source water protection programs, proper treatment, and distribution infrastructure O&M programs (Health Canada 2002), coupled with effective public health improvement and protection programs (Emde et al. 2001).

Once the balance and compromise between net gain in community health (e.g., physiological, psychological, economic, and social factors) relative to net economic development (key factors determining a community's social, economic, and psychological health) is breached, issues of water scarcity, unacceptable pollutant concentrations, and undesirable acute or chronic health effects become paramount. Adverse acute, or chronic, waterborne disease outbreaks or syndromes may occur. Individuals, especially the immunocompromised or immunosensitive, could become very ill or die. There is a net cost to individuals, their families, the health-care system, the community, and the environment. When individual and community health is compromised, the economic viability of that community is equally compromised.

Changes in regulatory requirements (e.g., improved filtration, improved chemical/microbial detection methods for newly regulated chemicals/waterborne pathogens) that mandate interaction between water utilities and health agencies to correct performance, usually require different levels and types of communication and/or collaboration than may have existed previously. Carried out properly, this approach minimizes conflict, eliminates adversarial agency positions, and helps to minimize the public's fears and mitigate adverse community outcomes during the crisis or incident. Everyone ultimately benefits, but there is typically a rather high price to be paid—the short- and long-term costs associated with the outbreak and its aftermath. Acceptance of the different but complementary roles, responsibilities, and jurisdictions of all partners and agencies involved with safe drinking water benefits the entire community. Such acceptance also serves to protect the community environment and ensure the safety of its drinking water, and continues to protect the community's health and well-being (Emde et al. 2001). Territorialism, confrontation, and conflict work against these outcomes.

CHAPTER 6
FUTURE DIRECTIONS AND NEEDS

Karen M.E. Emde, Daniel W. Smith, James A. Talbot, Les Gammie, and Nelson Fok

INTRODUCTION

Advances in knowledge over the past two centuries have resulted in the ongoing evolution and revision of drinking water regulations and standards (Pontius and Clark 1999). New technology has allowed for more timely and cost-effective detection of very low levels of chemical and/or biological agents of concern and has identified emerging agents of public health concern. An improved understanding of the fate, survival, and persistence of chemical and/or biological contaminants has spurred advances in source water protection programs, water treatment processes, infrastructure design, O&M practices, and community health surveillance activities, and has ultimately improved the medical care administered to individuals exposed to waterborne contaminants for short or long periods of time (Bartram 1999; Cohn, Cox, and Berger 1999; Geldreich 1992). Appendix A summarizes the workshop discussions held in July 2002 where invited representatives from international water utilities, environmental and health regulatory agencies, consulting engineering firms, public health regulatory agencies, water and public health analytical laboratories, consulting engineering firms, and police departments, as well as public health consultants, academia, and physicians, gathered to discuss current and future issues and to identify needs to better track and characterize health risks caused by infrastructure failures.

LEARNING FROM OTHER INDUSTRIES

Many cases exist, within and outside of the water industry, that illustrate different approaches to establishing and maintaining effective and proactive interagency collaboration. These can be categorized as performance correction or performance improvement. Both approaches have resulted in new or improved technology and processes, problem-solving strategies, investigation protocols, standards and legislation, and have ultimately benefited the community as well as the industry.

Total Quality Management

"Total quality management" (TQM) represents a corporate philosophy, vision, and management strategy with the objective of creating an organization committed to customer satisfaction, continual process and product improvement, teamwork, and leadership (Bendell 1998). TQM is more than just having a quality control program. In a TQM model, everyone in the organization becomes responsible for continually fostering teamwork and ensuring quality, improvement, innovation, and customer satisfaction. TQM was developed as a result of an identified need to improve business competitiveness and market share. The principles of TQM have been incorporated into International Organization for Standardization (ISO) standards, such as ISO 9000 and ISO 14000 (ISO 2003b), ISO/TC224 (ISO 2004), and others, some of which created the foundation for local/regional/national standards and regulations for a product or service (ISO 2003a).

The TQM movement has had many experts, each of whom developed and applied their own interpretation of fundamental TQM principles. The founding father was W. Edwards

Deming, a physicist, mathematician, and statistician in the early 20th century (Bendell 1998). Deming has been credited as a leader in applying statistical quality control principles together with interpersonal management strategies to improve manufacturing and nonmanufacturing processes and organizations (ASQ 2003). Other TQM leaders include Armand V. Feigenbaum (TQM), Joseph Juran (the quality spiral), Kaoru Ishikawa (quality control circles), Shigeo Shingo (zero defects), and Philip B. Crosby (zero defects) (Bendell 1998).

There is no single set of steps that instantly implements and guarantees TQM for an organization. However, the simple, basic elements common to most TQM philosophies are as follows:

- Clearly define the vision and/or goal and then strive to improve existing conditions and/or create innovative ways to achieve the goal.
- Make the vision happen by striving to achieve long-term goals—rather than short-term, bottom-line-driven goals—and foster team pride of participation and ownership in the corporate vision.
- Work toward continual improvement by thoroughly understanding all factors that contribute to the success and/or failure of any single step of the process, and then eliminate those steps that cause and/or promote failure.
- Foster an environment conducive to continual learning and improvement of all levels of the workforce.
- Work as a team. Break down barriers that hinder teamwork.
- Ensure that the entire team is rewarded for their commitment and their work to accomplish the goals.
- Pay attention to the customer and understand what the customer expects from the organization and product.

Two different approaches to TQM are presented in the following sections, and are generically referred to as the performance correction model and the performance improvement model. There can be significant differences in approach and motivation when applying these models to a for-profit corporation (e.g., 3M Company, St. Paul, Minn.) as opposed to publicly owned corporations and organizations (e.g., public health departments, publicly owned water utilities, emergency response agencies). Fostering a climate of quality improvement, innovation, and invention in a profit-oriented company typically represents future products and profits. In contrast, practice of that same philosophy by publicly owned corporations and organizations has little guarantee of immediate, direct financial gain.

Water utilities are faced with a ceiling for water rates that may not fully cover all costs of quality improvements. Public health departments and first-response agencies are faced with the constraint of working within set annual budgets that must cover all responsibilities of the organization, not just drinking water. However, the long-term benefits of TQM to the organizations, as well as to the entire community, can be measured with respect to improved water quality, improved reliability and longevity of community water infrastructure, safe drinking water, reduction and/or elimination of exposure to waterborne contaminants, reduction and/or elimination of exposure to waterborne disease, improved community health protection programs, improved interagency cooperation and collaboration, and so forth.

Performance Correction Model

An example of a performance correction model is the Federal Aviation Administration (FAA) approach to the investigation of airline crashes or air incidents. The FAA is a division of the U.S. Department of Transportation (DOT) and is responsible for issuing and enforcing the regulations and minimum standards governing the manufacture, operation, and maintenance of aircraft; certifications for airmen and airports; investigation of major/significant accidents and incidents; and response to National Transportation Safety Board (NTSB) requirements (FAA 2003). The FAA must provide the NTSB with an agency plan of action that addresses each safety recommendation made by the NTSB within 90 days after the issue date (FAA 2003).

The FAA approach uses standardized protocols and procedures for investigating aircraft crashes or incidents throughout the United States and coordinates with civil aviation agencies outside of the United States. These investigation protocols use an accident investigation quality assurance program to improve the quality of future accident and incident investigations. The outcome of the investigation is to identify safety deficiencies and unsafe conditions, which are then referred to responsible program offices in the FAA for further evaluation and corrective actions. Trend analysis of all accident and incident data, as well as other safety data, is conducted to identify safety issues and patterns for the aviation industry. Finally, the FAA reports their findings to the NTSB.

The NTSB is an independent, federal agency with a mandate from the U.S. Congress to investigate every civil aviation accident as well as significant accidents in other modes of transport to determine the probable cause; to conduct their own special investigations and safety studies; and to issue safety recommendations to prevent future accidents or incidents (NTSB 2003). The NTSB uses FAA findings to develop and issue safety recommendations for the airline industry. The NTSB is not part of DOT or affiliated with any DOT agencies. The NTSB–FAA formalized reporting relationship has resulted in improved aircraft construction, design, and O&M, as well as more stringent personnel training requirements and work programs, better safety procedures, and enhanced inspection programs.

Many published reports of individual waterborne disease outbreak investigations and incident reports appear to follow the performance correction model just described. That is, actions were taken and relationships established after the outbreak was confirmed and after the investigation of cause and remedial action plans has been established. The actions taken are typically retrospective.

Individual local and provincial/state public health and environmental regulatory agencies and water utilities may choose to function similar to the FAA model, and may coordinate their activities at a local or regional level. Unlike the NTSB–FAA model, most countries do not have a single, central, independent government agency given the formal authority or responsibility to perform the following functions: conduct independent investigations, coordinate information about the causes for and consequences of significant/serious water treatment and delivery failures and waterborne disease outbreaks, and develop and issue safety recommendations targeted at preventing future incidents or outbreaks.

In the United States, environmental health has been identified as one of the most fragmented and poorly defined areas of public health (CDC 2002, Pew Environmental Health Commission 2001). This finding is not unique to the United States, but rather can be observed in many countries, especially during widely publicized, major, unexpected disease outbreaks. The recent responses and reactions to the worldwide severe acute respiratory syndrome (more commonly referred to as SARS) outbreak is an effective example of this issue.

There are many federal, provincial/state, and local departments and agencies with environmental mandates that also include a public health component (e.g., departments of environment, labor, agriculture, Indian affairs [also known as First Nations and/or Aboriginal Affairs in other countries], emergency services, defense, and national parks) other than the actual department of health or environment. Similar parallels can be observed with multiple agency/department expectations and requirements concerning the protection, production, supply, and delivery of water in the community.

Fragmentation of regulatory responsibilities, jurisdictions, and resources can create uncertainty about who is responsible, who should be contacted, or who could provide assistance to manage and/or control environmental health problems and issues (Pew Environmental Health Commission 2001, CDC 2002).

Changes in regulatory requirements (e.g., improved filtration, improved chemical/microbial detection methods for newly regulated chemicals/waterborne pathogens) that mandate interaction between water utilities and health agencies to correct performance usually require different levels and types of communication and/or collaboration than may have previously exsted. Carried out properly, this approach minimizes conflict, eliminates adversarial agency positions, and helps to minimize the public's fears and mitigate adverse community outcomes during a crisis or incident. Everyone ultimately benefits, but there is typically a rather steep price to be paid—the costs (short and long term) associated with the outbreak and its aftermath. The direct economic costs of the *E. coli* outbreak in Walkerton, Ont. (Canada) in 2002 have been estimated at Can$64.5 million (Livernois 2002).

Performance Improvement Model

The 3M Company (3M) illustrates a performance improvement model that fosters improvement and innovation in the absence of a demonstrated need for performance correction or remedial measures. For almost an entire century, 3M has encouraged creativity and given employees freedom to take risks and try new ideas (3M Company 2002). This has resulted in more than 50,000 innovative products, including sandpaper (1904), cellophane tape (1930), magnetic video tape (1954), sterile wound-closure adhesive strips (1960), disposable face masks for respiratory protection (1967), self-stick notes (1980), hot-melt fiber-optic connectors (1990), dry-view laser imaging systems (1994), microflex circuits (1998), and many others. This company also has a history of listening carefully to their customers, commercial partners, and consumers, which has lead to further research and development to create new, innovative products. 3M also has an active commitment to invest in its employees and the communities where they work, including support of environmental efforts and environmental sustainability, education, and social and economic development.

The performance enhancement model can occur when there is a strong, positive relationship established and an active desire and willingness to create improvements or innovations that will benefit all partners. The desire to create improvements or innovations is a proactive rather than reactive approach.

This performance enhancement model approach is being actively promoted by multiple agencies involved with producing and delivering drinking water (e.g., AwwaRF, AWWA, USEPA, many water utilities, infrastructure component manufacturers, consulting engineering firms), agencies involved with the protection of community health (e.g., Health Canada, CDC, Agency for Toxic Substances and Disease Registry, WHO, Pan American Health Organization, local and state/provincial health agencies), and academia (e.g., universities, technical schools, certification programs). There is a shared vision for improvement, rooted in past experiences and knowledge of what worked, what didn't work, and what could have been done better based on previous incidents of waterborne contaminant exposures and resulting health outcomes. The goal becomes the creation and successful implementation of the processes, in the absence of a demonstrated need for performance correction and remediation, which builds on the collective pool of knowledge and experience to create improved products and manufacturing/delivery processes, while ensuring the protection of the community's health and provision of safe drinking water.

Table 6.1 compares basic organizational needs and objectives that are used in developing community protection programs by water utilities, public health and regulatory agencies, and emergency/first-response agencies. All of these groups share the common objective of protecting their community, but each group approaches their segment based on the responsibilities and jurisdictions of their individual organization.

GAPS

For the purpose of this discussion, the word *gap* can be used to describe a breach in a barrier or to describe disparate knowledge based on differences in one of the following areas:

- Training and experience; for example, environmental engineering versus public health academic training programs
- Evaluation and interpretation of results and information, such as water quality versus community disease incidence, based on training and perspective
- Available versus needed technology for contaminant detection, quantification, and characterization
- Lack of knowledge or information about an event, an agent, a consequence, etc., such as a new or emerging agent, new mode of transmission, etc.

Due Diligence and Regulations

Due diligence is a concept of law that has been incorporated into many professional standards of conduct and practice. Due diligence can be thought of as

- A standard that guides conduct
- A duty for best practice
- A performance indicator
- Ultimately, a legal defense

Organizations, including water utilities, are governed by statutes and regulations, as well as duties mandated by common law. The duties of common law include contracts, fiduciary responsibilities, product liability, and negligence (Thomson and EnviroChex Associates 2002). There is also the concept of the "directing mind," which is the person (or persons) who has been given executive authority over the organization (Thomson and EnviroChex 2002). It becomes the responsibility of that person (or persons) to ensure that the organization is meeting the required

Table 6.1
Program development objectives

	Water utility	Public health agency	Regulatory agency	Emergency/first-response agency
Basic goal	Provide safe drinking water, water for firefighting, and water for sanitation, which are part of protecting a community's health; manage utility capital infrastructure and processes.	Protect community and individual health.	Protect community health by setting and enforcing applicable regulations and legislation.	Protect community (individuals, businesses, institutions, and infrastructure) from confirmed or suspected sources of harm.
Assets (as defined by organizational culture, language, and perspective)	Utility infrastructure; treatment processes; source and delivered water; trained utility personnel	Individuals in community; infrastructure for monitoring, detection, and treatment of disease in the community	Individuals in community, environment of community (e.g., air, water, land)	Individuals in community and entire infrastructure of community (e.g., businesses, buildings, ability to conduct business)
Basic Objectives in Program Development				
1. Asset management and protection	• Develop and implement criteria for infrastructure design, construction, operation, monitoring, maintenance, renewal, and emergency protection programs.	• Develop health promotion and education programs, infectious/notifiable and/or chronic disease surveillance programs. • Perform outbreak investigation. • Enforce public health legislation and regulations, emergency protection programs.	• Protect community health and environment by setting, monitoring, and enforcing regulations.	• Protect community, individuals and infrastructure by monitoring, investigating, and enforcing applicable legislation and regulations; and by securing emergency sites and coordinating with other agencies during emergency situations.
2. Product liability	• Develop and implement written SOPs for key utility processes and activities • Perform process control and regular monitoring and evaluation of critical product/process characteristics. • Develop and implement product quality standards and procedures to deal with deviations from standards.	• Develop and implement ongoing disease surveillance programs with evaluation of community patterns and trends. • Implement immunization programs. • Implement disease prevention programs. • Enforce legislation and regulations.	• Develop and implement ongoing water quality (source and delivered) surveillance programs and review results. • Monitor key environmental/ public health indicators with evaluation of trends and patterns. • Enforce legislation and regulations.	• Develop and implement written SOPs for carrying out key community protection and emergency response programs. • Monitor key community protection parameters with evaluation of trends and patterns. • Enforce applicable legislation and regulations.

(continued)

Table 6.1 (Continued)

	Water utility	Public health agency	Regulatory agency	Emergency/first-response agency
3. Organization liability	• Follow practices and principles of due diligence in design, operation, maintenance, management, evaluation, and renewal of infrastructure and treatment processes. • Report procedures to evaluate water quality results and communicate to the community. • Respond to customer concerns and complaints—evaluation of trends. • Respond to advances in applicable technology, critical new knowledge, and definitive community trends.	• Follow practices and principles of due diligence, or precautionary principles, in the design, operation, management, evaluation, and upgrading of key public health programs, even when there may be insufficient evidence available at a given point in time to establish a definitive cause–effect relationship for a suspected health risk. • Respond to advances in applicable technology, critical new knowledge, and definitive community trends.	• Follow practices and principles of due diligence, or precautionary principles, when designing and writing new legislation and regulations. • Respond to advances in applicable technology and critical new knowledge for revision of existing standards or setting new standards. • Respond to advances in applicable technology, critical new knowledge, and definitive community trends.	• Follow practices and principles of due diligence, or precautionary principles, in the design, operation, management, evaluation, and upgrading of key community protection programs, even when there may be insufficient evidence available at a given point in time to establish a definitive cause–effect relationship for a suspected community risk. • Respond to advances in applicable technology, critical new knowledge, and definitive community trends.

standard of care, which means implementing all steps needed to reasonably minimize the potential of breaching a statute (Thomson and EnviroChex 2002). Failure to meet the expected standard of care can result in liability against the organization, its officers, and employees. The gold standard for water utilities and health agencies is developing a mutual culture of excellence through education, training, leadership, and collaboration in order to go beyond the minimum requirements of regulations and due diligence.

Regulations set the minimum acceptable standards of care, which include operations, product quality, and professional conduct, and are based on best available knowledge at the time the regulations were written and passed into law. Regulations also consider the fair and equitable costs of implementation, and try not to discriminate or cause unjustifiable hardship for smaller water utilities and health agencies by setting unmanageably high minimum requirements. Regulations represent the minimum expected—and accepted—community standards, not the maximum. Regulations define what an organization or individual must or shall do; there is no discretionary, regulatory leeway applied when these minimum expectations are not achieved. The use of the words *may*, *should*, *ought to*, or other words of similar meaning and intent within an act and/or regulation defines those situations where there is opportunity given to exercise discretion.

Advances in knowledge and technology may be incorporated into existing or new regulations, but there is typically a lag period of several years for this to occur. Guidelines and standards allow for flexibility in creation and modification, and are easier to quickly update to incorporate applicable, new knowledge in engineering, science, medicine, and so forth.

The concept of best practice implies that prior knowledge of these advances, such as new and emerging pathogens/contaminants, routes of exposure, codes of practice, process technology, O&M technology, and so on, will require a level of change to be made in the daily routines and judgment processes of the organization. Thus, the definition and implementation of best practices and due diligence change as gaps in knowledge are bridged.

Gaps in Knowledge

Training

Protecting the community from adverse or harmful health consequences of exposure to waterborne contaminants, whether the result of failure to adequately protect source water quality, failure of treatment processes, or infrastructure failures, involves multiple community agencies and professional disciplines. These include, but are not limited to, public health protection, medicine; epidemiology, basic science (e.g., biology, chemistry, physics), environmental engineering, process engineering, construction engineering; economics, law, political science, and first-response agencies (e.g., police departments, fire departments, hospitals).

There are often significant differences in academic training programs and professional requirements for each discipline involved. Professional certification programs may require additional training and experience beyond academic credentials accepted as that profession's minimum accepted entrance standard. Requirements for recertification of licensed professionals (e.g., those in the engineering, medical, science, public health, and water industries) may not be mandatory after licensing but, rather, are left up to each certified individual in that profession or their employer. Requirements for cross-discipline training are usually the responsibility of the individual, but may be initiated and directed by job skill and knowledge requirements. Licensing requirements are typically on a provincial/state-by-state basis rather than federal. Some professions have allowed "grandfathering"—a factor that was described as one of the many significant, contributing factors to the 2000 Walkerton, Ont., *E. coli* O157:H7 outbreak (O'Conner 2002).

Ongoing training, cross-training (including interdisciplinary programs), and/or retraining of staff require that the agency or employer have sufficient resources of available staff and funds. No one has unlimited resources, and as such, organizations must make the best use of those resources available. To be effective, short- and long-term benefits for the organization must be demonstrated, as well as a firm commitment by management to fund, support, and promote these programs.

Educational outreach and/or cross-training programs can include workshops and seminars, actual courses at recognized educational institutes, in-house programs, distance education, and so on. A benefit to multidisciplinary educational outreach is the ability for personal contact and interaction with other professionals involved with safe drinking water, public health, and community protection. There is an ability to exchange practical experiences and information, an ability to learn about the perspectives of other groups involved, and, most importantly, an ability to form professional contacts for future interaction. These programs do not need to have elaborate technical agendas or exorbitant registration fees. By keeping costs low, a program that is focused on current issues and obtaining good speakers ensures that the outreach plan will be successful. Starting time and length of the workshop/seminar program can be a significant factor for out-of-town participants who may only be able to attend for one day.

Technology

Investigations into health effects resulting from infrastructure failures have often been hampered by the following issues:

- Unpredictable and transitory nature of most failure events
- Immediate hazards at the failure site such as ruptured pipelines, loss of electrical power, loss of physical access to the site, flooding, earth subsidence, etc.
- Timely access to the suspected cause
- No, or poor, reporting of health consequences from the population served by the failed infrastructure
- Limitations in contaminant detection technology
- Availability of adequate, reliable field samples to confirm or disprove causal relationships

As a result, the true extent of acute and chronic health effects caused by infrastructure failures is not yet well known, understood, or characterized. There are other environmental sources and exposures of potential contaminants not linked to water distribution system failures, including

- The clinical effects from waterborne exposures may be poorly differentiated from other possible causes of the symptoms of disease.
- The variable dose–response relationships observed among the exposed population
- The low reported incidence of significant, adverse community effects; endemic rates of notifiable community disease may not be known or well-characterized, making early detection of low-level spikes in disease rates difficult.
- Some waterborne contaminants may have long latency or incubation times, making association with a single (or series of) infrastructure failure event(s) complicated and opaque.

As with regulations, there are minimum accepted and expected standards of technology and analysis defined for safe drinking water, as well as identification and quantification of waterborne contaminants. These community/regional/federal expectations may be referenced by legislation, or they may be referenced by guidelines for those regulations. Multiple agencies, including local, regional, and/or federal, have set standards for acceptable analytical technology, as well as appropriate methods of sample collection, transport, analysis, reporting, and evaluation of findings. Examples include *Standard Methods for the Examination of Water and Wastewater* (APHA, AWWA, and WEF 2005), *Bacteriological Analytical Manual* (USFDA 2001), and *Official Methods of Analysis of AOAC International* (AOAC International 2002). These may not be uniform throughout all jurisdictions involved. There may be local and/or regional differences in criteria for acceptance, rejection, interpretation, reporting, and/or evaluation of the significance of the findings.

Technological and analytical standards are continually evolving. The catalysts for this evolution can be found in both the performance correction and performance improvement models. Table 6.2 summarizes technology gaps and needs for water utilities and health agencies. Further discussion of these needs is provided in appendix A (workshop summary).

Table 6.2
Gaps in utility and health agency technology for identifying and characterizing health risks caused by infrastructure failures

Gap	Needed by: Water utility	Needed by: Public health agency	Direction needed	Benefits
Inadequate, real-time, cost-effective, sensitive technologies to routinely monitor, identify, and characterize waterborne contaminants of potential risk to health in the community water system	Yes	Yes	• Targeted, applied research programs for priority waterborne contaminants to develop new technology or modify existing technology to meet these needs	• Faster detection of potential problems • Faster response times • More efficient use of resources
Inadequate, real-time, cost-effective, sensitive technologies to quickly link infrastructure failures to changes in health status of individuals/communities affected by that failure	Yes	Yes	• Targeted, applied research studies (i.e., water utility, health, others) • Targeted epidemiological studies • Targeted studies linking use of hydraulic models for prospective epidemiological prediction	• Faster detection of potential problems • Faster response times • More efficient use of resources • Improved annual and strategic planning
Inadequate, real-time, cost-effective, sensitive technologies for linking acute illnesses to waterborne contaminant exposures from infrastructure failure events	Yes	Yes	• Targeted, applied research studies (i.e., water utility, health, others) • Targeted epidemiological studies • Reassessment of known rates of acute waterborne diseases with respect to historical records of community infrastructure failures and contaminant identification • Targeted, applied dose–response studies	• Earlier detection of potential problems • Faster response times • Improved interventions to control and/or minimize health effects • Better use of resources • Improved strategic planning • Improved standards for waterborne contaminants
Inadequate, real-time, cost-effective, sensitive technologies for linking chronic illnesses to waterborne contaminant exposures from infrastructure failure events	Yes	Yes	• Targeted, applied research studies (i.e., water utility, health, others) • Targeted epidemiological studies • Reassessment of known rates of specific chronic diseases with respect to historical records of community infrastructure failures and contaminant identification • Targeted, applied dose–response studies	• Earlier detection of potential problems • Faster response times • Improved interventions to control and/or minimize health effects • Better use of resources • Improved strategic planning • Improved standards for waterborne contaminants

Communication

The process of communication involves the transmission of verbal or written information and/or ideas through the use of common, clear systems of symbols, language, visual/written signs, and/or behaviors. The workshop discussions for this project, supported by published literature on waterborne outbreaks and/or infrastructure failure investigations, validate the importance and need for effective, open, honest, and collaborative interpersonnel and interagency communications in order to mitigate and/or eliminate health risks caused by infrastructure failure (see chapters 2, 3, and 4, and appendix A). Chapters 2, 3, and 4 discuss ways to initiate and maintain communication processes with the various community stakeholder agencies and/or groups, water utilities, and the public.

Clarity, honesty, and openness in communication between all stakeholders become the critical elements for effective problem solving, system improvement, and creative innovation. It is the cornerstone for successful collaboration, as well as for creating a deeper understanding of the other groups' abilities, needs, and perspectives. When the description and/or information provided by a particular viewpoint is unclearly, or poorly, communicated, resulting discussions can resemble those described by the 19th century poet, John Godfrey Saxe, in his poem about six blind men describing an elephant (Saxe [N.d.]).

There are many examples of models that illustrate effective and productive communication strategies. One model is the stakeholder consultation model (Macfarlane and McIntosh 2003). This model is routinely used by governments and regulatory agencies for planning and setting new limits, policies, regulations, and guidelines; and by industry to improve products, delivery of services, and so forth (OGC 2003, Australian Food Secretariat 2000, USEPA 2003, Chartier and Gabler 2001). Stakeholder consultations can be quickly organized to deal with newly identified threats or challenges, such as the recent National Governors Association (NGA) summit for U.S. public health officials and homeland security directors to discuss critical policy and management issues related to state and national concerns about bioterrorism and emerging disease outbreaks (NGA 2003).

Stakeholder consultation models use tools such as targeted collaboration with established expert/technical/advisory committees or groups; targeted communication with specific stakeholder groups (local, regional, or international), public meetings or town-hall forums; and discussions at conferences, workshops, surveys, and so forth. If done properly, this process ensures improved transparency in decisions made, accountability for actions taken, increased stakeholder awareness/involvement/confidence, and a better understanding of stakeholder perspectives (Australian Food Secretariat 2000, Chartier and Gabler 2001, OGC 2003, Macfarlane and McIntosh 2003).

A second common model is the voluntary outreach, consultation, and collaboration model. This model has also been described as a quality circle approach, a focus group approach, or a team approach. This model has been used within organizations to improve product quality, working conditions, customer service, and so forth, and has also been used as part of extramural stakeholder programs to voluntarily, and collaboratively, develop and implement new standards, protocols, technologies, and so on. ISO uses this approach, as well as the stakeholder consultation model (Macfarlane and McIntosh 2003).

ISO is a nongovernmental entity that consists of a network of the national standards institutes of 146 member countries. Each member country has a single and equal vote. The members of ISO are part of their national government structures. ISO acts as a bridge between government interests, industry and business interests, and the broader interests of individuals and society as a whole (ISO 2003a). Adoption and implementation of ISO standards is voluntary, although many member countries have incorporated specific ISO standards into their national standards, regulations, and legislations. ISO standards have also been incorporated into standards and specifications of many industries and businesses throughout the world.

ISO has standards for products (e.g., standardization of screws, standard paper sizes); standards for performance and safety (e.g., equipment standards); standard protocols (e.g., standards for computer communication and interfaces, standard documents for transport of dangerous goods), standards for analytical methods; standards for quality management and quality assurance (ISO 9000); standards for environmental management systems (ISO 14000); and is developing standards for service activities related to drinking water supply and wastewater systems (ISO/TC224; ISO 2004). Developing and implementing common standards promotes trade and commerce, public safety, and protects public health and the environment on a local, regional, and global scale.

Each ISO member has an equal voice in the development of any standard, which may have special significance to their country's economy and well-being. Expert technical committees form the backbone of the ISO standards development processes. ISO develops standards that are specifically requested, and for which there is a defined market need. Standards are reviewed every 5 years to determine whether they should be maintained, updated, or withdrawn (ISO 2003b).

Both of these models, as well as others, can be implemented by water utilities and public health agencies (with involvement by other key, community stakeholder agencies, as appropriate) for activities such as the development of joint protocols for issuing boil-water advisories; emergency response procedures; proactive surveillance (health and water utility) for emerging waterborne contaminants; development of a common database that quickly and specifically identifies sites in the community that could be possible sources of waterborne hazards from an infrastructure failure event or, conversely, identify sites that may be especially vulnerable; development of policies and procedures that facilitate parallel tracking of specific water quality complaints with infrastructure failure events and reported community health outcomes; and so on.

CONCLUSION

The purpose of the workshop held for this project was to have water utilities; public health, environmental, and regulatory agencies; medical personnel; engineering disciplines; and first-response agencies mutually explore the needs and capabilities of detecting, preventing, and correcting adverse health outcomes caused by infrastructure failures. This requires independent, as well as joint, surveillance systems that are capable of quickly identifying, monitoring, and correcting infrastructure failure events. These systems should also be capable of identifying introduced waterborne contaminants and their zone(s) of influence and exposure(s), and identifying acute and/or chronic health consequences caused by failure events, together with monitoring and evaluation of these efforts to ensure successful interventions.

Communication, coordination, and collaboration among all of these groups is essential should there ever be a danger to health as a result of an infrastructure failure. Developing and furthering these relationships and building trust long before occurrence of the failure event ensures efficient and effective interventions. However, the real goal of this strategy is to build relationships while improving current practices and technologies. Cooperative and collaborative extramural efforts are vital to develop better, or new, detection and prevention programs and/or technology for management of water utility infrastructure while protecting the public and community health.

APPENDIX A
WORKSHOP SUMMARY

Karen M.E. Emde, Daniel W. Smith, James A. Talbot, Les Gammie, and Nelson Fok

Summary of Workshop to Develop Methods to Evaluate Health Risks Associated With Infrastructure Failure (Awwa Research Foundation Project No. 2773)

July 21–23, 2002, Edmonton, Alta., Canada

INTRODUCTION

The issue of health effects, acute and chronic, caused by distribution infrastructure failures are the concern and responsibility of the local water utility, the public health department, community medical care providers, and, in special circumstances, emergency/first-response agencies. It is the responsibility of the water utility to provide safe drinking water, as mandated by regulations and legislation governing the treatment and distribution of potable water. This includes maintenance and renewal of the distribution infrastructure in the community. It is the responsibility of the engineering community to properly design fully functional water systems and to update the design as new technology is recognized. It is the responsibility of public health agencies to prevent disease and promote health in the community. It is the responsibility of medical care providers to treat disease. Finally, it is the responsibility of emergency/first-response agencies (i.e., police department, fire department, ambulance services, and others) to respond to actual or perceived emergency situations in the community, as well as to contain the emergency situation and work in conjunction with other community agencies having specific jurisdiction. In a suspected or actual emergency situation, the first-response agency, usually police and fire department personnel, will act to contain and handle the emergency at hand, deal with public fears and rumors, be involved as part of an initial investigative team, and may be part of the final investigation, especially if the emergency is the direct or suspected result of a deliberate violation of criminal legislation. Hospital emergency room staff (i.e., physicians, nurses, aides, and others) should also be consulted, kept informed, and trained in the community emergency strategy, as they will be the first individuals who will treat mass casualties.

Ideally, the investigation of waterborne disease, from all possible causes, is most effective when performed by collaborative teams composed of public health agency and water utility personnel. Actively fostering an early liaison and collaboration with community first-response agencies (i.e., police, fire, ambulance, and others) will benefit the overall strategy developed to protect that community. Each group brings responsibilities, insights, and expertise that collectively result in enhanced activities, and thereby minimizing duplication of efforts. Active acceptance of the different, yet complementary roles and responsibilities of all of the partners involved in safe drinking water and protection of the public's health results in a net benefit for the entire community.

Table A.1
Workshop objectives

1. Develop details of methods that could be used to better assess potential or known health risks resulting from the failure of the municipal distribution system infrastructure. These tools would allow water utilities, public health agencies, and primary care facilities to pro-actively respond to infrastructure failure situations and make more efficient use of all available resources while effectively protecting the public's health.
2. Develop recommendations for the best methods, or combination thereof, to initiate enhanced programs for water utilities and public health agencies.
3. Develop recommendations to identify gaps in current health risk assessment guidelines with respect to health risks from distribution system infrastructure failures, and ways to effectively bridge or minimize those gaps.
4. Develop recommendations to help small, mid-sized, and large water utilities to prioritize infrastructure maintenance, and repair and renewal programs to minimize the potential for adverse community health effects.
5. Develop recommendations to help water utilities and public health agencies to more quickly and effectively cope with distribution infrastructure failure due to natural and/or human-made disasters for a given water utility.
6. Develop recommendations to identify areas of health risk concern from distribution system failures that require future research efforts to better characterize the potential for adverse health effects, but more importantly to allow water utilities and health agencies to refine existing strategies to minimize this risk potential to the public.

WORKSHOP STRUCTURE

Objectives and Participants

The purpose of this workshop was to convene an invited, international group of experts representing the interests and perspectives of water utilities, public health agencies, regulatory agencies, and first-response agencies to discuss the issues of health effects, suspected or confirmed, resulting from failure of distribution system infrastructure. The objectives of this workshop are presented in Table A.1. This group was given the task of trying to develop a method, or methods, to more efficiently detect and evaluate health risks due to infrastructure failure. The participants reflected the spectrum of agencies involved with these issues. A list of participants is presented in Table A.2.

Workshop Presentations

Six expert speakers were invited to present information on infrastructure failure and vulnerability, natural and human-made disasters, application of distribution system models to assess distribution system vulnerability, intervention strategies to reduce the risk for waterborne disease outbreaks, process and health risk assessment methods, and disease surveillance.

Table A.2
Workshop participants

Participant	Organizational affiliation
Ms. Susan Ancel, P. Eng	EPCOR Water Services Inc., Edmonton, Alta., Canada
Dr. Jeff Aramini	Health Canada, Guelph, Ont., Canada
Mrs. Edie Ashton	Alberta Provincial Laboratory of Public Health, Edmonton, Alta., Canada
Dr. M. Anita Barry, M.D.	Boston Public Health Commission—Communicable Disease Control, Boston, Mass., United States
Mr. Philip Broad, P.E.	Sydney Water, Sydney, Australia
Dr. Don Bursill	Cooperative Centre for Water Quality and Treatment, Salisbury, Australia
Detective Mike Cook	Edmonton Police Services, Edmonton Alta., Canada
Ms. Gabrielle Deagle, P. Eng.	EPCOR Water Services Inc., Edmonton, Alta., Canada
Dr. Rolf Deininger	University of Michigan, Ann Arbor, Mich., United States
Ms. Karen Emde	University of Alberta, Edmonton, Alta., Canada
Mr. Nelson Fok, CIPHI	Capital Health, Edmonton, Alta., Canada
Mr. William Foley, P.E.	Boston Water and Sewer Commission, Boston, Mass., United States
Dr. Les Gammie	EPCOR Water Services Inc, Edmonton, Alta., Canada
Mr. Edwin Geldreich	E. Geldreich Consulting Microbiologist, Cincinnati, Ohio, United States
Dr. Walter Grayman, P.E.	W.M. Grayman Consulting Engineer, Cincinnati, Ohio, United States
Mr. David Hartman, P.E.	Cincinnati Department of Water Works, Cincinnati, Ohio, United States
Mr. David Hiltebrand, P.E.	AH Environmental Consultants Inc., Hampton, Va., United States
Dr. Steve Hrudey, P. Eng	University of Alberta, Edmonton, Alta., Canada
Mr. Pat Lang, P. Eng.	Alberta Environment, Edmonton, Alta., Canada
Dr. Deborah Levy	Centers for Disease Control and Prevention, Atlanta, Ga., United States
Dr. David Lipsky	City of New York Department of Environmental Protection, New York, N.Y., United States
Mr. Morris Maslia	Agency for Toxic Substances and Disease Registry, Atlanta, Ga., United States
Mr. Kevin McLeod, CIPHI	Alberta Health, Edmonton, Alta., Canada
Mr. Wayne Miller, P.E.	Windsor Utilities Commission, Windsor, Ont., Canada
Mr. David Muto	Seattle Public Utilities, Seattle, Wash., United States
Mr. Keith Osborn	United Utilities plc., Warrington, Great Britain
Mr. Kenneth Rotert	U.S. Environmental Protection Agency, Washington, D.C., United States
Dr. Martha Sinclair	Monash University Medical School and Cooperative Centre for Water Quality and Treatment, Melbourne, Australia
Dr. Daniel W. Smith, P. Eng	University of Alberta, Edmonton, Alta., Canada
Mr. Clark Svrcek	University of Alberta, Edmonton, Alta., Canada
Dr. James A. Talbot, M.D.	James Talbot Prof. Corp., Edmonton, Alta., Canada
Mr. Ryan Ulrich	Awwa Research Foundation, Denver, Colo., United States
Mr. Richard Walker, P.E.	Western Australia Water Corp., Leederville, Australia
Ms. Brenda Woo	Health Canada, Edmonton, Alta., Canada

Infrastructure Failure and Vulnerability (Presented by David Hiltebrand)

This presentation focused on the vulnerability of the distribution system and consequences of natural disasters, operation and maintenance (O&M) practices, distribution system design, and deliberate disruptions on water plant and distribution operations and failure. Natural and/or human-made disasters have been shown to result in additional infrastructure failure, such as disrupted electrical services, road/bridge blockage or damage, subsidence of soil, and contamination of water supplies. Water utilities not only have to deal with the immediate impacts of the disaster, but there is often a period of significant cleanup and repair that disrupts community water services. One major source of difficulty for distribution system operations during a failure event is that accurate, accessible maps of the distribution system are often not available, are outdated, or have been destroyed or damaged during the disaster.

Additional information was presented on distribution system failures caused by system design and operation. Consequences of this category of failure are typically noted as a water quality concern, such as biofilm, corrosion products, tastes and odors, discolored water, increased turbidity, and so forth. Benefits of routine mains cleaning to control these problems were presented. Issues of excessive water storage (tanks, mains) and poor or inadequate mixing were presented as a commonly occurring category of infrastructure failure, with suggestions for remedial action. This included eliminating excessive or unused storage capacity, increasing turnover, and proper calibration of instrumentation. Benefits of backflow-prevention and cross-connection control programs were presented as significant means to protect the distribution system from intrusion.

The issue of intentional contamination of the distribution system was presented using three case studies. The first was an actual case where there had been a threat of an unknown chemical added upstream of a water supply. Retrospective review of procedures used during this suspected emergency illustrated the lack of an effective response plan (i.e., no personal protective equipment was used during sampling, no special sampling handling protocols or tests were ordered, and there was no specialized assistance available). The other two cases represented false alarms, but also illustrated similar deficiencies in preparedness response.

Natural and Human-Made Disasters: Health Risks due to Failure of the Water Supply System (Presented by Rolf Deininger)

Information was presented to further illustrate the vulnerability of water distribution systems to externally induced failure. Five case studies of waterborne disease outbreaks were presented where human-made influences, which were in some cases coupled with extreme weather, resulted in failure of the water distribution system. The presentation discussed the need for real-time monitoring ability that is quick, easy to use, and cost-effective. Improved instrumentation and protocols would allow water utilities to be proactive rather than reactive when an emergency situation exists. This would also help to protect public health and prevent cases of community waterborne disease. Cleanup and disinfection of distribution systems after a disaster has occurred are extremely time consuming, costly, and may require procedures and testing that are not part of the routine protocols of the water utility.

Water Distribution System Vulnerability: Application of Distribution System Models (Presented by Walter Grayman)

Case studies were presented that illustrated the use of hydraulic distribution models to track the movement of contaminants introduced into the distribution system. The case studies illustrated the value of hydraulic models as forensic tools for investigation of the source of contaminants, tracking contaminants (distance, time), and tracking community exposure potential. As hydraulic models used by many water utilities are skeletonized, further development and calibration of an existing hydraulic model would be needed to be effective for real-time contaminant monitoring.

An overview of the types of threats to the distribution system (e.g., physical disruption, biological or chemical contamination, loss of consumer confidence) and the potential impacts and entry points for contaminants was presented. The utility and cost benefits of instrumentation for inline monitoring for contaminants was discussed, as were research and development needs for improved, real-time monitoring technology.

Need for Better Intervention Strategies to Reduce the Risk for Waterborne Outbreaks (Presented by Edwin Geldreich)

This presentation focused on the entire water system, with the distribution system as the final barrier to prevent entry of contaminants to the water supply. The point emphasized was that there is no single barrier that can consistently inactivate all waterborne pathogens or not be overwhelmed by pollution events. Case studies of 27 communities were presented to illustrate key points of this presentation. The need for a multiple-barrier approach that operates on a 24/7 schedule, with routine monitoring of process and performance, was reinforced.

Weaknesses were discussed in response to waterborne disease outbreaks, including poor monitoring, lack of appropriate response to positive or suspicious laboratory results, and poor interpretation of significance of routine laboratory data as the significant factors in waterborne disease outbreaks. Also addressed was the need for interagency collaboration in the review and setting of new guidelines and legislation, emergency and incident planning, and handling of public concerns.

Process and Health Risk Assessment Methods: Expectations and Realities (Presented by Steve Hrudey)

This presentation described key elements and definitions of risk, risk analysis, and application of risk analysis. Differences between the process risk and the health risk assessments were described using case studies and illustrative examples. The processes used to define risk, identify hazards, quantify risk and hazards, and estimate probability of occurrence and consequence were outlined.

Health risk processes were shown to be significantly different than process risk assessments. Health risks are a function of exposure and health response, both of which can be short- or long-term phenomena. Health risk assessment processes and information required for acute and chronic exposures and health outcomes were compared and contrasted. The difference between risk and safety was presented, which is significant when setting regulatory guidelines for contaminants, assessing exposure and routes of exposure, and establishing causality. The difference between prediction of risk and validation of outcome was defined, with the point emphasized that there is usually very little opportunity to validate the risk assessment. The limitations of risk assessment were also presented, a key point being that the risk assessment is only as good as the evidence/data that is available.

Disease Surveillance—A Missing Cog in the Wheel of Water Supply Management (Presented by Jeff Aramini)

The epidemiology of several recent waterborne disease outbreaks (e.g., Walkerton, Ont.; North Battleford, Sask.) were discussed, as well as a prospective epidemiological study for the link between waterborne disease cases and turbidity fluctuations. These cases were used to illustrate the limitations of current disease surveillance tools and the ability for real-time monitoring of waterborne disease.

Information about the future challenges for epidemiological monitoring was presented, which included the development of methods to more quickly and effectively monitor and evaluate disease trends to detect clusters of disease in the community. Also discussed were various types of surveillance tools, including monitoring of pharmaceutical sales, emergency room visits, calls to health hot lines, school absenteeism surveillance, and others. The need for timely surveillance data to quickly identify waterborne disease events was presented as one key element of a comprehensive strategy to control the incidence and spread of community waterborne disease.

Breakout Group Structure and Function

Breakout groups convened to discuss the daily theme. Participants were randomly assigned to one of two groups, which focused on the distribution system (i.e., engineering) or community health (i.e., public health). The theme for day 1 was assessment of hazards (distribution system and community health); day 2 was intervention strategies (distribution system and community health); and day 3 was future needs and research in distribution system issues and community health impacts. One third of the participants in any given breakout group was reassigned to a new group on a daily basis. This was to ensure that all participants had ample opportunities to interact during the course of the workshop, as well as to share their expertise and experiences. Participants recorded their responses to theme questions and presented those to the group for discussion at each respective session. Two additional recorders were provided per group.

Workshop Results

Analysis of Vulnerability

A distribution system's vulnerability to failure was determined to be a function of several key factors:

1. Engineering features including
   - Criteria used for distribution system design, including system pressure and demand (e.g., fire flow versus water quality)
   - Actual system design and construction

- Location of distribution system (e.g., geology, climate, proximity to other buried structures such as sewer lines, proximity to other sources of contaminants)
- Design for security of distribution system

2. System O&M practices including
   - Routine O&M program (e.g., execution of program, frequency, evaluation, leak detection program, cross-connection control program, backflow-prevention program)
   - Presence and/or absence of process control
   - Distribution infrastructure monitoring program, including monitoring changes in water quality throughout the system
   - Operation of system for maximum rather than minimum security and protection of water treatment processes and infrastructure components (e.g., covered, secure water storage reservoirs)
   - Written standard operating procedures
   - Active incident and emergency plan
3. Training issues including
   - Staff education (entry level, retraining, ongoing training, cross-training)
   - Staff morale and interest in job
   - Staff attitude (e.g., interest in simply meeting minimum legislated standards versus interest in performing at a level that exceeds legislated standards)
   - Understanding of changes to legislated requirements (e.g., understanding of sources and nature of hazards and potential waterborne threats to public health, understanding of improvements for control and/or elimination of potential waterborne threat)
   - Ability and willingness to communicate (oral and written) with other water utility staff and other key community agencies (e.g., public health, regulators, consumers)
4. Factors beyond control of water utility such as
   - Natural disasters (e.g., earthquakes, floods, violent weather),
   - Human-made disasters (e.g., human error, vandalism, sabotage, terrorism, acts of war)
   - Public fears, rumors, and customer expectations

Vulnerability of the community to waterborne disease (acute and chronic) as a result of infrastructure failure included the following elements:

1. Individual health factors such as
   - Age and immune status
   - Presence/absence of preexisting, acute health condition (e.g., recent surgery, recent chemotherapy, other infectious disease)
   - Presence/absence of preexisting, chronic health condition (e.g., cancer, human immunodeficiency virus, diabetes)
2. Community and/or cultural factors such as
   - Language (e.g., poor or no fluency in oral/written language of location/country)
   - Restricted movement within the community (e.g., self-imposed, imposed by others [i.e., institutionalized populations], imposed by circumstance [i.e., poverty, lack of mobility, etc.])
   - Proximity to water treatment plant (WTP), type of WTP processes, type of distribution system, and O&M of distribution system

3. Perceived community vulnerability due to
   - Rumor and fear factors (individual, as well as the community as a whole)
   - Perception and prioritization of individual rights and privileges versus needs and rights of the community
   - Individual and community knowledge of hazards (e.g., sources of hazards, increased knowledge resulting in changed expectations and demand for improvements)
   - Ability to activate alternative strategies during an infrastructure failure (e.g., availability of alternate potable water supply, ability to temporarily relocate)

Another vulnerable community group was identified, that is, commercial and industrial customers. They depend on potable water as a critical component of a manufactured product or the manufacturing process, or where access to potable water supports key equipment (e.g., air-conditioning, cooling systems) or critical business operations (e.g., entertainment events, airports, hotels). Industrial vulnerability to the consequences of infrastructure failure was determined to be a function of the following elements:

1. Direct use of unaltered distribution water in manufacturing and/or products resulting in
   - Human exposure potential
   - Event exposure potential
   - Critical product exposure potential, possibly causing the product to serve as a secondary vehicle to expose the community to waterborne hazards
2. Dependence on minimum critical flow of potable water to support manufacturing/ business processes, equipment, and products
3. Emergency requirements for fire protection

Analysis of Potential Hazards and Risk

Potential hazards resulting from distribution infrastructure failure, irrespective of cause of failure or origin of hazard, were identified as follows:

- Introduction of microorganisms (e.g., bacteria, protozoans, viruses) and macroorganisms (e.g., worms, vermin) as contaminants, increased biofilm formation potential possibly resulting in adverse ecological shift in biofilm populations (e.g., pathogen survival and persistence in biofilm)
- Introduction of chemical contaminants (i.e., external to distribution system, formed within distribution system, etc.)
- Introduction of physical contaminants (e.g., soil, rocks, construction debris)
- Introduction of radioactive contaminants (e.g., due to geological factors in surrounding soil, waste radioactive products)
- Lack of water and/or intermittent water supply for hygiene, health, manufacturing, and fire protection
- Psychological hazards (e.g., fear, stress, or anxiety for community residents, water treatment and distribution staff, and industrial and commercial users)

- Secondary hazards (e.g., burns, scalds, fires) resulting from temporary measures needed during potable water supply disruption
- Actual damage to infrastructure components and subsequent diversion of resources (e.g., staff, time, equipment, money) needed for repair and additional monitoring
- Increased potential for liability (e.g., corporate, personal, civil) and increased insurance costs
- Increased potential for adverse personal, corporate, or political reputation and its consequences (e.g., job loss, job change, loss of customer confidence and loyalty, customers switching to alternate sources of potable water)
- Increased financial costs, directly (e.g., cost of alternative water supplies, costs of repair, rate increases) and indirectly (e.g., lost time due to illness, temporary plant slowdown/shutdown, product recall) as a result of the failure

Discussions about health risks from distribution infrastructure failures determined that critical gaps for health risk analysis included

- Lack of information available about health risks associated with various categories of distribution system failure (i.e., confirmed waterborne outbreaks/disease incidence related to the failure event, relatively small number of reported cases, what type and magnitude of distribution failure will result in community/neighborhood health effects, etc.)
- Inadequate information available linking chronic health conditions to waterborne contaminant exposure caused by a failure event (i.e., confirmed cases related to a specific failure event, long lag time between exposure and onset of confirmed symptoms and diagnosis of health condition, etc.)
- Inadequate information about the true rates of endemic versus epidemic disease in the community that hinder the ability to detect the onset of new disease outbreaks
- Inadequate, quick, cost-effective detection technology and monitoring (utility and public health surveillance) capability for all potential categories of distribution system health risks
- Inadequate or incomplete health and utility baseline data that correlate with each other for failure events and the threshold levels that trigger further, more intensive investigation
- Failure to communicate information to other community agencies that may need to know (i.e., interagency communication may range from friendly and cooperative to adversarial to nonexistent, or one agency may not believe that the information is of significance at the time of occurrence, etc.)
- Lack of clear boundaries and jurisdictions, as well as changing standards and expectations, between responsibilities of the water utility to provide potable water and the public health agency to protect the health of the community (e.g., water quality regulations are based on the best available knowledge at the time that the regulation was drafted, but the time lag between setting the standard and acquiring new knowledge about hazards and health effects is immense)
- Separating water treatment process failures from actual distribution system failures (e.g., failure at the WTP will impact the distribution system and may or may not result in distribution infrastructure failure and/or altered distribution water quality)

During the course of the workshop, it was acknowledged that the true extent of health effects caused by various categories of infrastructure failure was not well known or characterized. It was determined that existing detection and surveillance methods (water utility, public health agency, medical provider, etc.) were not adequate at this point in time to truly quantify the extent of the health effects in the community.

There were discussions about sample collection and handling (i.e., what to collect; how much to collect; chain of custody, transport, handling, storage, and analysis as a function of real or perceived hazards associated with the contents of the sample; samples being handled, stored and treated as "evidence" of an infrastructure failure event and consequences of those types of samples versus samples collected for routine monitoring, etc.). Examples were presented of suspected deliberate contamination of water supplies by extremely toxic chemicals, wherein those samples were shipped on a commercial passenger aircraft to an analytical laboratory without appropriate precautionary labeling, thereby creating the possibility of secondary and tertiary exposure to suspected hazards in the water sample container as a result of handling and transport. Routine strategies used by police departments in collecting, handling, storing, and maintaining chain-of-custody samples that could potentially be used as evidence may be of value to water utilities and public health agencies in investigating possible health effects due to infrastructure failures. Water samples collected under these circumstances must be considered and evaluated in a much different light than routine monitoring samples. These become the critical evidence needed to prove or disprove an association to drinking water contamination from the infrastructure failure event as well as between acute or chronic health effects due to exposure to hazards present in the contaminated water for a very small window of time.

Intervention Strategies

During workshop discussions, there were examples presented of differences in viewpoint, culture, and language between water utilities and public health agencies when describing health risks caused by infrastructure failures. At times, these diverging viewpoints were primarily the result of differences in training, professional culture, and language used. There was a final consensus reached that described the need for improved, enhanced, ongoing communication between all of the agencies involved with community waterborne disease and safe drinking water. This fundamental working relationship forms the foundation for intervention and prevention strategies for all agencies involved in waterborne disease detection and prevention.

The need for improved public health surveillance capability (i.e., enhanced surveillance systems rather than passive surveillance, using multifunctional surveillance programs and tools to track multiple community disease syndromes rather than only a single disease condition, establishing community rates of endemic disease, etc.) was confirmed to be very desirable. Similarly, the need for improved water utility surveillance (i.e., distribution system condition, water quality complaint recording and analysis of trends, etc.) was also thought to be highly desirable. Both types of programs may require additional resources than those currently available (e.g., staff, funding, facilities, perceived need, and benefits for these programs for the community and/or agency, management cooperation and support for programs, sharing of common resources and/or information).

The need for changes to training programs was also discussed. Currently, many of the professional discipline and operator training programs provide cursory information in non-core subjects. For example, public health training programs often do not have extensive course requirements in operation of WTPs and water distribution systems. Similarly, many environmental engineering and technical water staff training programs may provide a brief overview of waterborne disease agents and surveillance. There may be limited, or no, opportunities for cross-training at educational institutions because of the specified requirements for diploma/degree programs, lack of staff, lack of time, and lack of resources. The opportunities for on-the-job training and retraining often are limited by time, budget, and staff constraints. Lack of, or limited, contact with other professionals dealing with waterborne disease prevention and detection further hinders opportunities for increased learning and understanding of common issues.

Surveillance and monitoring tools and parameters that are currently used by many water utilities and public health agencies can provide valuable information if the information is subject to ongoing analysis and evaluation rather than merely collected for the sole purpose of meeting regulatory requirements and then filed. Concomitant evaluation of community water utility and public health data can provide valuable information about trends that may be missed when routine sampling is performed by a single agency.

Effective communication was determined to be a key element for all monitoring, surveillance, and incident response strategies. This involves communications among all stakeholder agencies (i.e., water, public health, medical health, first response, political, and others) involved in the investigation of health effects due to infrastructure failures. More importantly, this also involves communicating with the public. These efforts will realize short- and long-term benefit to the community.

The aftermath of an outbreak typically results in massive public fear and rumor; treating, comforting, and compensating the victims; seeking and punishing the "guilty"; and implementing corrective actions and measures to prevent future, similar occurrences (e.g., new and stricter legislation and regulations, new or improved regulatory agencies with different or more comprehensive enforcement tools, more surveillance). As time passes (e.g., post-Milwaukee [MacKenzie et al. 1994], post-Gideon, post-Walkerton,) and there is a program of remedial measures in place (e.g., stronger legislation, more monitoring, more surveillance), the challenge becomes to maintain that vigilance. These programs are expensive in terms of financial and human resources—if the improved efforts result in decreased or no detectible community waterborne disease, they may not be maintained in the long term. Public opinion and political agendas tend to be fickle—they often respond to the short-term emergency, not the long-term gain. The absence of, or decrease in, cases may well be the result of increased monitoring and surveillance efforts. However, the absence of cases may also be viewed as a reason to cut back or discontinue a particular program. It may not be seen in light of the success of the enhanced surveillance being conducted by all stakeholder agencies.

The challenge becomes to move all of these systems closer together to develop a common, early warning system for detection and prevention of community disease and water quality problems. This requires dedicated efforts by all agencies involved to collaborate, cooperate, and share knowledge and information. A joint strategy for waterborne disease prevention requires different alert levels that trigger interagency communication to investigate the cause for concern. This requires institutional procedures in place to ensure that the process will actually work and that there is a commitment by all those involved to try and make the process succeed. This requires a common, clearly defined understanding of what the problems are, including

- Case definition for public health versus case definition for water utility surveillance and monitoring
- Evidence that the problems represent an actual deviation from an established community baseline and are potentially linked to water rather than other sources of exposure
- Description of the agents involved and an understanding of their methods of operation (e.g., infection, intoxication, acute versus chronic health effects)
- Description of the nature of the failure event
- Time of the event
- Location of the problem
- An understanding of the global/community implications of the problem (e.g., number and types of customers potentially exposed or affected; acute versus chronic effects.)

There were many discussions about the future potential for common geographic information systems (GISs), that is, mapping health surveillance data (i.e., patient information, confirmed cases, pharmaceutical sales data, hospital visits, etc.) on the same grid as water quality, water complaints, and distribution system failure data. There are numerous drawbacks that prevent immediate implementation of this approach. First is the issue of patient confidentiality with respect to access to health data. Access to medical information is strictly protected by provincial/state and federal legislation to protect the rights of the individual. Use of this information would require development of measures to ensure that patient confidentiality and privacy is incorporated into a collaborative GIS surveillance program. The second issue is the presence, or absence, of a collegial, collaborative, open working relationship between the community agencies and the water utility. There are additional issues concerning

- Maintenance of the GIS (which agency or all agencies?)
- Evaluation and interpretation of data and patterns, and the significance to the water utility and to public health agencies
- Distribution system information currently on utility GISs (i.e., skeletonized versus detailed, validated hydraulic model, etc.)
- Quick and easy access to information and by whom
- Compatible computer systems
- Funding
- Staff training
- Long-term staff and management commitments

Smaller agencies or municipalities may not have these resources or the capability of implementing this type of collaborative surveillance. However, there may be opportunities for partnering with larger agencies, if all parties are willing. Again, issues of resource allocation (i.e., staff, time, funds, equipment, etc.); a clear understanding of the needs, issues, and benefits to all partners; and the nature of the relationship (e.g., open and collegial, adversarial, nonexistent) between the larger and smaller agencies would need to be addressed first.

Future Needs

Throughout the workshop discussions, there were many examples of future needs and research expressed by all of the participants. These can be categorized into the following areas:

1. Epidemiological studies to
   - Provide direct evidence of health effects (e.g., acute and chronic, and to what extent these can be observed in the exposed population) caused by different categories of distribution system failures
   - Provide guidelines to more accurately predict waterborne disease as a function of a distribution system failure event and patterns of community (to help provide guidance to public health agencies and water utilities, which triggers agency response levels)
   - Establish causes of endemic disease in the community and break down the attributable fraction of waterborne disease into the proper components
2. Laboratory and field studies to
   - Develop improved, rapid, cost-effective diagnostic tests for specific agents (microbial and chemical)
   - Develop improved, cost-effective, rapid indicator systems and real-time testing technology
   - Develop a better understanding of dose–response relationships
   - Develop a better understanding of survival, persistence, and transmission of waterborne contaminants in the distribution system that may result in adverse health effects
3. Utility studies to
   - Better understand the relationships between infrastructure failure events and community health outcomes
   - Better understand the consequences (acute and chronic) for the utility and the community of different categories of failure event
   - Develop alternative O&M strategies to minimize or prevent future infrastructure failure events
   - Link the use of hydraulic models for better tracking of sources of contaminants, propagation of contaminants, and exposure potential for prospective rather than retrospective surveillance
4. Relationship and communication strategies to
   - Overcome the barriers in reporting events to other community agencies that need to be informed of infrastructure failure events
   - Improve communication between all community agencies involved in protection of public health and safe drinking water
   - Improve the use of technology to promote interagency communication and collaboration

5. Interagency collaborative strategies to
   - Improve tracking systems
   - Improve monitoring and sampling (i.e., evidence-gathering techniques and evidence storing)
   - Develop improved strategies to protect agency front-line staff (e.g., utility, health, first responders) who are the first to deal with the failure event and the potential contaminants
   - Develop cross-training strategies so that agencies with different mandates on a common community issue can better understand each other's perspective and develop better collaborative relationships and programs

APPENDIX B
LITERATURE REVIEW

Karen M.E. Emde, Daniel W. Smith, and James A. Talbot

INTRODUCTION

Drinking water distribution systems, municipal and private, are ubiquitous in our society, and many consumers and industries served by such systems make automatic and implicit assumptions about the safety and aesthetic quality of water being delivered (Emde et al. 2001). Distribution system operation and maintenance is one of the cornerstones in the delivery of safe drinking water (Ellison et al. 2001, Pierson et al. 2001, Kirmeyer et al. 2000, Antoun et al. 1995).

The U.S. Environmental Protection Agency (USEPA) has specified the most significant area for future utility needs and spending: renewal and rehabilitation of the distribution infrastructure (USEPA 1997, Davies et al. 1997). Water distribution systems represent a significant capital and operational investment for water utilities, with as much as 80 percent of the total utility expenditure devoted to maintenance and renewal of distribution and transmission mains (Deb et al. 1997). Utilities are faced with increasing costs for maintenance and operations, which need to be weighed against the costs associated with water quality deterioration, contaminant intrusion, and the potential for community acute, or chronic, waterborne disease as a result of contaminant exposure. More and more often, water utilities are being held liable for adverse effects in their community, such as flooding, property damage due to soil subsidence in the vicinity of a leak or main break, community waterborne disease outbreaks, or exposure to chemicals caused by distribution infrastructure failure (Ellison et al. 2001; Sylvester, Reiter, and Waldron 2000).

Water forms part of the critical infrastructure of the community and, as such, must be properly maintained and protected (Luthy 2001, Hebert 2001, Grigg 2000, PAHO 1998). Failure of any single component of the community-critical infrastructure affects the functional integrity of other critical infrastructure present in that community. There are three generic failure scenarios:

1. A disruption where one critical component failure results in disruption of a second critical community infrastructure (e.g., power failure causes water distribution pumps to fail, causing interrupted water supply)
2. A disruption where one critical component failure provokes an independent disruption of a second critical infrastructure in the community (e.g., failure/disruption of the transportation system results in inadequate or no supplies being delivered to the water plant, resulting in interrupted water treatment and distribution)
3. Failure of critical municipal infrastructure components (e.g., natural or human-made disaster) as a result of a common cause (Peerenboom 2002)

Figure B.1 illustrates the interrelationship of drinking water with other community-critical infrastructures. Disruption to any, or all, components affects the ability of the water utility to provide an adequate quantity and quality of potable water to the residents of the community. The magnitude, nature, persistence, and ultimate impact of the disturbance and the condition of the infrastructure will influence the degree to which water supply is interrupted and water quality is changed. During the actual disturbance event, water flow in adjacent pipe-loop systems is altered, thus temporarily changing delivered water quantity and quality in affected mains.

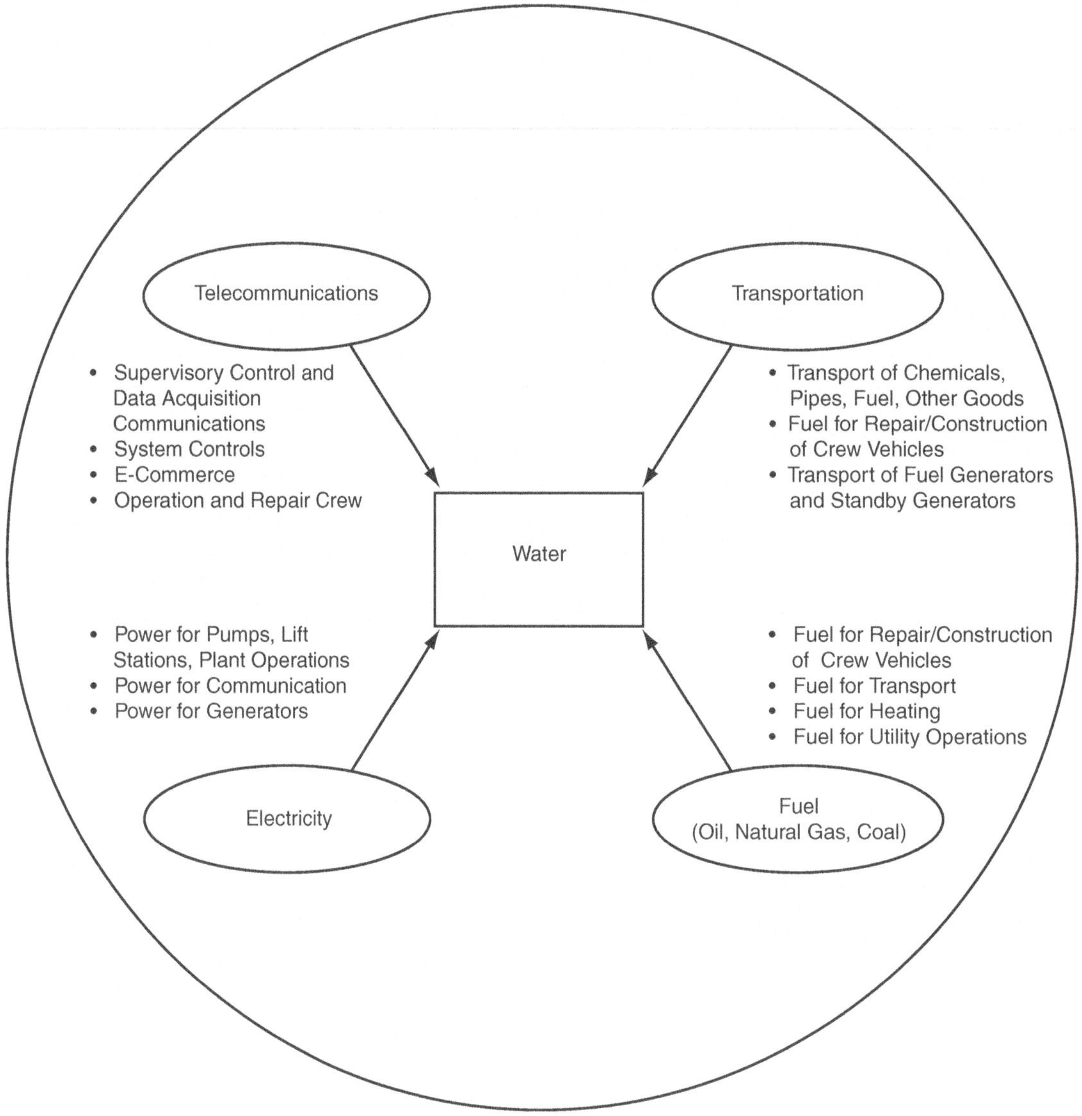

Figure B.1 Interdependence of drinking water on other critical community infrastructures

Protection of community drinking water is needed not only from the effects of catastrophic events such as violent weather, earthquakes, or deliberate acts of sabotage but also from everyday events. Events such as main breaks, pressure reductions, disruptions caused by routine construction, cross-connections, human error, or other potential causes of system failure can result in intrusion of biological and/or chemical contaminants into the distribution system. These types of situations are commonplace and routinely handled by every water utility. However, stakeholder agencies may or may not be informed of every single incident, which makes tracking of potential short- and long-term health effects of introduced contaminants difficult or impossible.

Preventing exposure to waterborne contaminants from drinking water and protecting public health is the result of water utility vigilance in monitoring and maintaining the water treatment process, but, more importantly, the municipal water distribution system. Water utility efforts are accompanied by the coordinated and sustained efforts of stakeholder agencies, such as public health authorities, environmental regulators, and others. In its most ideal form, this interagency partnership is collegial, dynamic, open, and active.

DISTRIBUTION SYSTEM PROPERTIES THAT CREATE VULNERABILITY FOR FAILURE

Distribution System Materials

Water distribution systems contain multiple individual components such as pipes (distribution/transmission/service lines), valves (line/pressure regulating/control valves), meters, fire hydrants, water reservoirs, standpipes, storage basins, and pumping stations (Kirmeyer et al. 2000). The engineering design criteria for an individual community and projected community water demands dictates the type, size, number, and installation methods used for these materials (Cesario 1995).

Individual components can be subject to materials defects, and poor or improper transport, storage, handling, installation, and maintenance, any of which can influence the reliability, performance, and life span of that part (Burlingame et al. 1998; Kirmeyer, Wagner, and Leroy 1996; Vik et al. 1996). Improper handling of new pipe materials, coupled with inadequate postconstruction disinfection, can result in bacteriological and chemical contamination of the pipe being put into service (Geldreich 1994). Poor installation and workmanship can result in dissimilar metals being joined together, creating corrosion cells and shortening the performance life of the pipe (Bradford 1993). Stringent specifications for distribution system components, reliable suppliers, internal material specifications, and a materials testing program of components is only part of an overall strategy that can be used by a water utility to manage their distribution system (Burlingame et al. 1998). Material deficiency together with engineering design errors and poor construction can cause structures to fail, which happened to a newly constructed water tower in Fredericton, N.B., Canada (Dawe, Seah, and Abdel-Zaher 1993).

Corrosion of Distribution System Components

Water distribution systems can be subject to corrosion of external and internal surfaces, resulting in pipe deterioration, main breaks, and altered potable water quality and quantity. External corrosion is the primary cause for structural failure of the distribution system, typically resulting in pipe failures and leaks (Ljunggren 1990). External corrosion is a function of the soil type (i.e., moisture, alkalinity, acidity, permeability of water and air, oxygen, salts, stray currents, and microorganisms) and the pipe material being buried (Fontana 1987). The most commonly used methods to control external corrosion are pipe coatings and cathodic protection (Jones 1992). An alternate strategy used by some utilities is to replace metal distribution and transmission mains with polymeric pipe materials (e.g., polyvinyl chloride [PVC], polyethylene [PE]) (Jones 1992).

Internal corrosion can also be the cause of structural failure in the distribution system but, more importantly, is a mechanism to significantly alter delivered water quality (Kirmeyer, Wagner, and Leroy 1996; Schock 1999). Impacts of internal corrosion of the distribution system include

- Introduction of chemical contaminants (i.e., lead, copper, zinc, cadmium, asbestos fibers, iron, and others) that may pose a potential health hazard
- Introduction of microbial contaminants, including direct and opportunistic pathogens, that can result in a potential health hazard
- General water quality deterioration in the form of altered color (e.g., green, blue, black, and red water) and/or taste (e.g., metallic, bitter, sulfide) causing consumer concerns about the aesthetic quality
- Increased pumping costs as a result of increased hydraulic friction due to internal pipe tuberculation and deterioration
- Increased costs of chemical treatment to control corrosion and maintain the proper chlorine residual throughout the system
- Deterioration of actual distribution system components and customer plumbing components
- Water losses due to main breaks and leaks
- Adverse impacts on wastewater collection systems, effluent, and sludge due to elevated metal concentrations in potable water from deteriorating distribution mains entering the wastewater collection system (Kirmeyer, Wagner, and Leroy 1996; Schock 1999)

Microbially mediated or influenced corrosion (MIC) is a term used by the corrosion industry to describe electrochemical processes catalyzed or facilitated by the growth of microorganisms (Borenstein 1994). MIC is not a new phenomenon, nor has it been recently discovered. The earliest reports of MIC in potable water systems were documented in 1912 and were associated with microbial deterioration of water reservoir linings (Schoenen and Schoeler 1985). Many other case reports exist, the majority involving some form of deterioration of the actual distribution system material or its components, causing release of corrosion products and organisms to the passing water phase. In some cases, this deterioration was linked to outbreaks of waterborne illness (Schoenen and Schoeler 1985; Emde, Smith, and Albritton 1994).

There are several ways in which distribution system microflora may influence the corrosion process. Certain microorganisms can influence corrosion by metabolism of corrosion inhibitors or degradation of protective coatings (Costerton and Boivin 1991, Ford and Mitchell 1990). These coatings are used to prevent internal corrosion of water mains. Examples of coating materials commonly used in water systems and approved by the American Water Works Association (AWWA) include vinyl, chlorinated rubber, epoxy, asphalt, and coal tar (Alber, Bruchet, and Shpirt 1989). Degradation of pipe coatings, whether by chemical or microbial means, has caused concerns about degradation by-products entering the water phase. Examples of degradation by-products include

- Methylketones and toluene from vinyl coatings
- Xylenes, C_8–C_{10} alkyl esters, and alcohols from chlorinated rubber
- Methyl isobutylketone and xylenes from epoxy coatings
- C_2–C_4 alkyl benzenes from asphalt
- Polycyclic aromatic hydrocarbons from coal tar coatings (Alber, Bruchet, and Shpirt 1989)

Studies that evaluate leaching from these coatings often do not consider the effects of attached microbial growth and subsequent deterioration of the coating in the water distribution system.

The regrowth of microorganisms within the distribution system, especially coliforms, has received considerable attention from the water industry. In contrast, the role of MIC and resulting water quality and distribution system deterioration has received relatively little attention. There have been many utility-directed research efforts with the purpose of reducing levels of organic carbon and other nutrients in treated water supplies for control of bacterial regrowth (LeChevallier 1990; LeChevallier, Olson, and McFeters 1990; van der Kooij 1990, 1988). Regrowth studies do not consider the effects of hydraulic shear on the existing internal biofilm development and ecology in the water distribution systems.

Geomicrobiological processes, such as the oxidation and reduction of metals, elements, and ions, function in the same manner, whether in water distribution systems or soils (Ehrlich 1990). Many microorganisms involved in geomicrobiological cycles are routinely found in water. Some form part of the resident flora of distribution system water and pipe biofilms (Emde and Smith 1992). Standard methods used for analysis of potable water may not be sufficient to routinely detect and enumerate these organisms. Standard methods were developed to monitor regulated indicator organisms rather than routinely monitor biofilm populations and levels in water distribution systems (Emde and Smith 1992).

Water utilities have several options to deal with internal distribution system corrosion. These include

- Alter the water quality entering the distribution system so that it is less corrosive (e.g., pH control, alkalinity adjustment, etc).
- Use corrosion-resistant linings and paints for the interior pipe surface.
- Use corrosion-inhibiting chemicals (e.g., phosphate corrosion inhibitors, sodium silicate).
- Use distribution materials that are less easily corroded (e.g., PVC and PE pipes).
- Change the design of the water system in new or renovated areas of the community to avoid low flow or dead ends in the system (Deb et al. 2000; Schock 1999; Kirmeyer, Wagner, and Leroy 1996).

Hydraulic Transients in the Distribution System

Several phenomena can result in hydraulic transients in the distribution system. Hydraulic transients can be rapid (e.g., water hammer, main breaks) or slow (i.e., events creating periods of low pressure). Common examples of hydraulic transients include

- Main breaks and water leaks
- Water hammer
- Temporary low water pressure due to events such as mains flushing/pigging, fire flow, shutdown of individual distribution system components (e.g., pumps), etc.
- Nonutility hydrant use (e.g., street cleaning, fire flow, valve checking, tank cleaning, unauthorized use) (Deb et al. 2000, Kirmeyer et al. 2000, Walski and Lutes 1994)

Main Breaks and Distribution System Leaks

Main breaks can be the result of many internal and external factors and forces acting on distribution pipes and can be classified into one of the following types:

- Emergency breaks
- Nonemergency leaks (AWWA 1986a)

An emergency break is typically the rupture of a transmission or distribution main or a severe leak in an individual customer service line (AWWA 1986a). A water leak of this magnitude can cause a serious pressure drop in the affected service area. This can result in intrusion of chemical and microbial contaminants by backsiphonage or cross-connections to the municipal system. Table B.1 summarizes incidents of distribution system contamination by this mechanism. In the majority of cases presented in Table B.1, mains rupture was the pivotal event that allowed entry of contaminants from other points in the distribution system (PNWS–AWWA 1996). In the majority of cases, health effects were neither noted nor tracked (PNWS–AWWA 1996). Other similar incidents have clearly illustrated the serious health effects that can result, such as chemical burns, infectious disease, and, in the most severe case, death (PNWS–AWWA 1996).

An emergency main break can disrupt service to the community and cause the following events:

- Flooding
- Low distribution system pressure resulting in backsiphonage
- Insufficient water for firefighting
- Insufficient or inadequate water for community health and hygiene
- Serious disruption to local business activities

In August 2000, an extensively corroded, 48-in. (1.2-m) water main broke in Kansas City, Mo., resulting in a 2-day boil-water order and serious disruption of downtown businesses (*The Business Journal* 2002). Thousands of employees were sent home, air conditioning at a major downtown hotel was disrupted, hotel guests were relocated, and new hotel guests were turned away (*The Business Journal* 2002). Kansas City had experienced earlier main breaks in 2000, primarily because of structural failure of the distribution pipe and rupture of the water mains by contractors working in the area (*The Business Journal* 2002). Similarly, two main breaks in New Port Richey, Fla., caused a total service disruption to 350 local homes (*St. Petersburg Times* 2001). During the repair, water utility crews damaged an underground power line, which resulted in a power outage in the community and prolonged water service disruption (*St. Petersburg Times* 2001).

A nonemergency leak is considered to be a known low-magnitude water leak or unknown water leak in the distribution system (AWWA 1986a). A utility leak detection program, water audit program, or water loss program (AWWA 1986a) typically detects these. Distribution system leaks can occur at valves, fittings, or in pinpoint holes in the distribution pipes. Water leaks can result in erosion of the pipe bedding material, ultimately resulting in the potential for structural failure of the main (Deb et al. 2000). Leaks can also augment the potential for increased external mains corrosion due to elevated moisture levels in the soil (Deb et al. 2000). Distribution system leaks may provide avenues for intrusion of soilborne biological and chemical contaminants during

Table B.1
Summary of municipal distribution failures due to backflow/backsiphonage and cross-connections, from 1923 to 1995

Year and location	Type of failure	Cause of failure	Contaminant	Community impact	Health impact
1923, Everett, Wash.	River water contamination of municipal distribution supply	Open bypass valve from unused lumber mill connected to mains; pressure drop in municipal system	Typhoid bacteria	Typhoid outbreak; lawsuit settled in favor of plaintiff	2 deaths; community typhoid outbreak
1936, Winooski, Vt.	River water contamination of municipal distribution supply	Open bypass valve from unused woolens mill used as fire booster supply water (river); pressure drop in municipal system	Typhoid bacteria	Lawsuit settled in favor of plaintiff	1 death
1942, Newton, Kan.	Sewage contamination of municipal distribution water supply	One of town's water supplies out of service; open hydrant froze; neighboring toilet sewage overflowed into hydrant box; no disinfection of mains put back into service	Enteric pathogens	Community gastroenteritis outbreak	More than 2,500 people ill
1947, Milford, Neb.	River water contamination of municipal distribution supply	Open connecting valve at pump house after a fire allowed river water to enter municipal supply	Enteric pathogens	Community gastroenteritis outbreak	150 people ill
1965, Madera, Calif.	Sewage contamination of municipal water wells	Poorly protected wellshed; connections between valve excavation and well pit; gopher holes in vicinity of well; sewage landfarming	Enteric pathogens	Community gastroenteritis outbreak	246 people ill; 2 hospitalized
1967, Renton, Wash.	Gasoline contamination of municipal distribution supply	Pressure testing of gasoline line using direct connection to municipal supply	Gasoline (2,000 gal [9,092 L])	Explosion; gasoline in municipal supply	1 person injured in explosion
1967, New England states	River water contamination of municipal supply	Fire service water (river water) mistaken for municipal supply had bubbler connected for drinking water	Hepatitis A virus	Community outbreak	7 persons ill
1969, Peoria, Ariz.	Arsenate-based herbicide backsiphonage into municipal distribution supply	Weed killer tank filled with garden hose; main break caused pressure drop and backsiphonage from tank	Arsenate-based herbicide (arsenic levels at 98 ppm; 0.02 ppm after 24-hour flushing)	Community water ban until arsenic levels <0.01 ppm	None noted
1969, Worcester, Mass.	Cross-connection between feces-contaminated sprinkler boxes and municipal distribution supply	Potable water connection to sprinkler boxes; local fire caused pressure drop resulting in backsiphonage from sprinkler boxes	Hepatitis A virus	Community outbreak; local football games cancelled for most of season	83 football team members and coaching staff ill

(continued)

Table B.1 (Continued)

Year and location	Type of failure	Cause of failure	Contaminant	Community impact	Health impact
1969, New Haven, Conn.	Backsiphonage of irrigation water into locker room drinking water supply	Municipal water main repair created siphon on supply line, resulting in backsiphonage of irrigation water standing around sprinkler heads	Hepatitis A virus	Community outbreak	More than 50 persons ill
1970, Cincinnati, Ohio	Backsiphonage of wine into municipal distribution system	Distilling tanks operated at higher pressure than municipal mains during fermentation; no backflow devices	Sparkling burgundy wine	Altered taste in municipal water	None noted
1971, Spokane Valley, Wash.	Backsiphonage of stagnant water into municipal distribution supply	Abandoned water well supplied an abandoned building connected to municipal supply; street washers filling from hydrant caused pressure drop in municipal supply	Coliforms	Coliform-positive samples at local restaurant	None noted
1972, Huachuca City, Ariz.	Backsiphonage of chlordane into municipal distribution supply	Exterminating company mixed pesticides using hose connected to municipal supply; company at higher elevation than municipal system	Chlordane (0.8 to 236 ppm)	Taste and odor complaints	1 person ill
1973, Marlboro Township, N.J.	Backsiphonage of chlordane into municipal distribution supply	Exterminating company mixed pesticides using hose connected to municipal supply; 24-in. (61-cm) main break caused backsiphonage through hose and into municipal system	Chlordane	Extensive mains flushing	None noted
1973, Woburn, Mass.	Backsiphonage of fungicide into municipal distribution supply	Faulty backflow valve in greenhouse; greenhouse chemical application system pressure was 300 psi [2,068 kPa] while city water pressure was 60 psi [414 kPa]	Fungicide	Odor complaints	None noted
1974, Soldiers Grove, Ill.	Herbicide contamination of municipal distribution supply	Replacement of fire hydrant with no valve in hydrant lead and poor valving in main, thus water reservoir shut down and distribution system drained resulting in backsiphonage from a tank containing herbicide	Balan (Elanco Products Co., Indianapolis, Ind.)	Flushing and rechlorination of municipal system	None noted

(continued)

Table B.1 (Continued)

Year and location	Type of failure	Cause of failure	Contaminant	Community impact	Health impact
1974, North Carolina	Boiler water additive contamination of municipal distribution supply	Faulty check valve at fertilizer plant allowed chemicals in boiler to enter municipal water supply	Boiler treatment chemical	Fast food restaurant soft drink contamination by boiler treatment chemical	Unknown number ill (more than 300 persons served soft drinks before problem found)
1976, Wenatchee, Wash.	Pesticide contamination of municipal distribution supply	Pesticide tank at higher elevation than town well; herbicide applicators filling trucks (pesticide and water) caused reservoir pressure drop and backsiphonage of pesticide into municipal system	Endrin (>130 ppb in initial samples once problem reported 2 days after incident occurred)	Families on same water line exposed (dermal and ingestion exposure)	Unknown
1976, Winnipeg, Man.	Fertilizer contamination of municipal distribution supply	Snow-clearing knocked over two fire hydrants; pressure drop in municipal system; backsiphonage of blue-colored fertilizer from local florist	Plant fertilizer	Customer complaints of blue water	None noted
1979, Chattanooga, Tenn.	Insecticide contamination of municipal distribution supply	Variable water pressure in municipal system due to multiple main breaks. A single house on municipal line suspected source that had been treated for termites; exterminator used tap water to dilute insecticide	Chlordane (>1,000 ppm)	Three blocks of contaminated water mains replaced; 47 homes vacated; payment of damages (total >$1,000,000)	4 persons with probable chlordane poisoning; 32 persons (minimum) exposed
1979, Kulm, N.D.	Herbicide contamination of municipal distribution supply	Negative pressure in water system; individuals using garden hoses to dilute DDT solutions in sprayers; backsiphonage to municipal distribution system	DDT	Customer complaints of taste and odor; extensive municipal flushing program	Temporary burning sensation on lips and throat of persons drinking water; minor gastrointestinal upsets
1979, Meridian, Idaho	Stagnant water contamination of municipal distribution supply	Routine hydrant flushing, system pressure drop; backsiphonage of stagnant water from local supermarket fire sprinkler system	*Clonothrix fusa* and *Zoogleora ramigera*	Customer complaints of taste and odor (onions)	None noted

(continued)

Table B.1 (Continued)

Year and location	Type of failure	Cause of failure	Contaminant	Community impact	Health impact
1979, Seattle, Wash.	Contaminated water entry into municipal distribution supply	Cross-connection between car wash reclaim water tanks (wash and rinse water) and municipal water line; restoration of high-pressure car-wash pump forced contents of tanks into municipal water lines	Brown, soapy water	Color complaints	None noted
1979, Roanoke, Va.	Insecticide contamination of municipal distribution supply	Water main repair; system pressure drop; exterminator treating nearby homes with chlordane for termite control; back-siphonage of chlordane into municipal system	Chlordane	Customer complaints of taste and odor; extensive municipal flushing program; litigation in favor of plaintiff ($13,000,000)	None noted
1979, San Antonio, Texas	Steam contamination of municipal distribution supply	Main break repair; water pressure drop; steam from heating boiler entered municipal main	Steam	None	None noted
1980, San Antonio, Texas	Chemical contamination of municipal distribution supply	Main break resulting in system pressure drop; private customer toilet tank additive backsiphonage into municipal distribution supply	Toilet tank additive	Blue water complaints	None noted
1980, Clallam County, Wash.	Detergent contamination of municipal distribution supply	Main break resulting in system pressure drop; backsiphonage of detergent solution from steam-cleaning tank	K-90 detergent	Green water complaint	None noted
1981, South Bend, Ind.	Natural gas contamination of municipal distribution supply	Galvanic corrosion between gas main and copper water service (house) resulted in natural gas entering municipal supply	Natural gas	Community complaints of sputtering gas appliances, frozen gas regulators, homes without heat, natural gas in water	None noted
1981, Norfolk, Va.	Salt water contamination of municipal distribution supply	Freezing rupture of backflow valve in shipyard saltwater fire protection system; positive pressure in saltwater system, thus contaminating municipal system	Sea water	Customer complaints of salty water	None noted

(continued)

Table B.1 (Continued)

Year and location	Type of failure	Cause of failure	Contaminant	Community impact	Health impact
1981, Santa Ana, Calif.	Oil contamination of municipal distribution supply	Incorrect plumbing of water-cooled hydraulic press; oil pressure higher than domestic water pressure; oil pumped into municipal water supply	Oil	Mains flushing to remove oil	None noted
1982, Shiawassee County, Mich.	Herbicide contamination of municipal distribution supply	Fire hydrant replacement; isolating valve nonfunctional; village distribution system shut down but not everyone contacted. System pressure drop; customer used aspirator-type sprayer to apply lawn herbicide; backsiphonage to municipal system	Malathion	Unknown	None noted
1982, New England states	Propane contamination of municipal distribution supply	Purging of propane tank at propane plant using municipal water. Propane tank pressure at 85 to 90 psi [586 to 621 kPa] while city water pressure at 65 to 70 psi [448 to 483 kPa].	Propane	Hundreds of people evacuated from homes; fires in two homes; one home gutted by propane explosion from toilet	None noted
1982, Bangor, Maine	Antifreeze contamination of municipal distribution supply	Water shut off to home to make street main repairs; system pressure drop; back-siphonage of antifreeze from home heating system to municipal main	Antifreeze	Green water complaint	None noted
1983, Woodsboro, Md.	Herbicide contamination of municipal distribution supply	Town pump failure resulting in system pressure drop; open gate valve between herbicide holding tank and town water system, resulting in contamination of water supply. Restoration of water pressure spread herbicide in town supply	Paraquat	Extensive flushing; sampling program; provision of alternate water supply	None noted
1984, Seattle, Wash.	Corrosion-inhibiting chemical contamination of municipal distribution supply	Failure of air-conditioning recirculation pump; water-to-waste air-conditioning loop temporarily set up; bypass of backflow prevention device in building; back-siphonage of water to municipal water supply	Nitrate–borate corrosion inhibitor	Customer complaints; extensive local flushing program	None noted

(continued)

Table B.1 (Continued)

Year and location	Type of failure	Cause of failure	Contaminant	Community impact	Health impact
1984, City of Manhattan, Kan.	Anticorrosive chemical contamination of municipal supply	Repair of water main break caused local system pressure drop; backflow of boiler water with anticorrosive chemical to drinking water at local elementary school	Anticorrosive chemical	Backflow prevention program initiated at county schools	None noted
1984, Macon, Ga.	Creosote contamination of municipal supply	Repair of private hydrant required water supply to be shut off; pumping of creosote continued with a connection to a fire service line; resuming municipal water service caused backsiphonage of creosote to municipal line	Creosote	Customer complaints about water taste	None noted
1985, Arpelar, Okla.	Pesticide contamination of municipal supply	Hose left inside tank containing pesticide mixture; main break caused pressure drop resulting in backsiphonage of chemicals to municipal system	Chlordane, malathion, Sevin (Bayer Crop Science, Research Triangle Park, N.C.), diazinon	Extensive flushing program for more than 2 months (cost approximately $100,000)	None noted
1985, Sacramento, Calif.	Pesticide contamination of municipal supply	Pump for chemical feed left on; backflow devices failed; main break in municipal system, resulting in backsiphonage of pesticide to municipal system	Malathion	Extensive flushing program; news warning for consumers to not consume water	None noted
1985, Grand Prairie, Texas	Xylene and ethylbenzene contamination of municipal supply	Suspected failed backflow devices; source of transmission to municipal system unknown	Xylene (118 ppm), ethylbenzene (17 ppm)	Extensive flushing program; 5 food processors in area had contaminated product, which was destroyed	None noted
1986, Withrow, Wash.	Herbicide contamination of municipal supply	Farmer filling spray tank to dilute 2,4-D; pressure drop in municipal system caused backsiphonage of tank contents to municipal system	2-4-D	Extensive flushing; community without water supply for 4 days	None noted
1986, Lacey's Chapel, Ala.	Sodium hydroxide contamination of municipal supply	Water was being added to tanker truck containing sodium hydroxide, municipal 8-in. (20-cm) water main break resulted in pressure drop and backsiphonage of tank contents to municipal system	Sodium hydroxide	Extensive flushing program; approximately 60 homes in area affected	A number of people received chemical burns after washing or showering

(continued)

Table B.1 (Continued)

Year and location	Type of failure	Cause of failure	Contaminant	Community impact	Health impact
1986, Fayetteville, N.C.	Pesticide contamination of municipal supply	Water used to fill pest control company truck; municipal main break resulted in pressure drop; backsiphonage of truck contents through building and into municipal system	Chlordane, heptachlor	Customer complaints of odor	None noted
1987, Fair Lawn, N.J.	Pesticide contamination of municipal supply	Water used to clean pest control company truck; municipal main break resulted in pressure drop; backsiphonage of truck contents into municipal system	Heptachlor, chlordane, Dursban (Dow AgroSciences, Indianapolis, Ind.)	63 homes and businesses without water for several days; extensive flushing program	None noted
1987, Gridley, Kan.	Pesticide contamination of municipal supply	Water used to clean feed store pesticide storage tank; municipal main break resulted in pressure drop; backsiphonage of truck contents into municipal system	Lexone DF (Dupont, Newark, Del.)	Community warning to not use water	None noted
1988, Cleveland, Ohio	Water soluble oil plus chemical additives contamination of municipal supply	Backflow valve on oil storage tank failed when tank pressurized, resulting in backflow of tank contents to municipal system	Oil, nitro-butyl morpholine, ethylniotrotrimethylene	Customer complaints of taste and odor; neigh-borhood flushing initiated	Several complaints of nausea and vomiting
1989, Fordyce, Ark.	Propane contamination of municipal supply	Accidental cross-connection between pressurized railroad tank car containing propane and municipal system	Propane	Two homes destroyed by propane-induced fire; local business seriously damaged; other homes had explosions of varying degrees when household taps/toilets used	Three injuries from explosions
1990, Tucumcari, N.M.	Industrial chemical contamination of municipal supply	Illegal connection to fire auxiliary line connected to municipal supply at plant; valve left open when plant operations resumed; backsiphonage of chemicals into municipal system	Toluene, phenol, benzene ethanol, nonanoic acid, decanoic acid, octinol, octanoic acid, heptanoic acid, butanoic acid, silicon, diconic acid, trihalomethane	Extensive community flushing program	3 people ill (nausea, vomiting) after drinking water

(continued)

Table B.1 (Continued)

Year and location	Type of failure	Cause of failure	Contaminant	Community impact	Health impact
1991, Missouri	Trichloroethane contamination of municipal supply	Cross-connection at local newspaper allowed backsiphonage of chemical into municipal system; disorganized municipal flushing spread contaminant throughout entire system	1-1-1-trichloroethane (420 µg/L)	Extensive community flushing program	None noted
1992, Ritzville, Wash.	Algal contamination of municipal supply	New municipal reservoir had single 16-in. (41-cm) line that served as water inlet and outlet; 12-in. (30-cm) reservoir drain line connected to 16-in. (41-cm) line and algal growth in drain sump. When water pumped into reservoir, venturi effect past the drain connection caused backsiphonage from drain sump into reservoir.	Blue-green algae	Customer taste and odor complaints	None noted
1994, Franklin, Neb.	Chlorodifluoromethane gas contamination of municipal supply	Repair of building air-conditioning system; a hole in coil of unit was route for chlorodifluoromethane gas (pumped at 1,000 psi [6,895 kPa]) to enter water circulating in air conditioner, which was connected to municipal supply; ultimately, chlorodifluoromethane gas entered municipal distribution system.	Chlorodifluoromethane gas	Customer complaints of taste and burning sensation in mouth; residents without water for 6 days during municipal flushing efforts	None noted
1995, New Roads, La.	Herbicide contamination of municipal supply	Farmer diluting herbicides in 300-gal [1,364 L] spray tank with municipal water; work crew accidentally cut water line, causing pressure drop and backsiphonage of herbicide into municipal system	Paraquat dichloride, atrazine	Customer complaints of taste, odor, and color; class-action lawsuit against water utility	Some persons had nausea, acute stomach pain, diarrhea, shortness of breath, lip sores, burning gums, leg cramps

Source: Adapted from PNWS–AWWA 1996.

episodes of low pressure in the distribution system (Moser 2000, Kirmeyer et al. 2001, Pierson et al. 2001).

Water Hammer

Water hammer is a positive-pressure wave that is created by sudden, large, positive-pressure changes in the distribution system and can travel up the pipe at the speed of sound and may be great enough to cause significant pipe and/or valve damage or total failure (Roberson and Crowe 1993, Geldreich 1996, Deb et al. 2000, Burns et al. 2001). Events such as rapidly opening and closing distribution system valves and fire hydrants are common causes of water hammer in a municipal distribution system. Water hammer can result in significant hydraulic shearing of pipe biofilms, releasing biofilm organisms, biofilm/corrosion sediments, and accumulated chemicals to the water phase.

Water hammer can cause serious failure of a building's plumbing systems and components. A severe water hammer in the fire protection system at the Washington, D.C., nuclear plant caused catastrophic failure of a large, cast-iron valve as the result of a false fire alarm (Arastu et al. 1999). There was serious flooding in some areas of the nuclear plant before remedial action and repairs could be started (Arastu et al. 1999). Water hammer has been suspected and confirmed in several waterborne disease outbreaks including *Legionella pneumophila* infection of patients in a tertiary-care hospital (Joly et al. 1993, Joly and Alary 1993). Failure of the hospital shock absorbers at individual sinks resulted in infection of susceptible individuals from aerosols generated during localized water hammer events (Joly et al. 1993, Joly and Alary 1993). Patients inhaling these aerosols became infected with waterborne, nosocomial *L. pneumophila*.

Temporary Low Water Pressure

Periodic reductions in distribution system pressure can result from a variety of causes including

- Periods of community peak water flow and inadequately sized mains
- Mains flushing programs
- Pipe leaks and pipe ruptures
- Distribution valve malfunction
- Closing of valves in sections of the distribution system
- Opening and closing of fire hydrants
- Sudden high water demands in other sections of the distribution system
- Periodic operation of equipment connected to municipal water lines (e.g., pumps, pressure washers) (Walski and Lutes 1994, Kirmeyer et al. 2001)

Temporary low water pressure can result in backsiphonage from industrial and private service connections or cross-connections to the distribution system, allowing intrusion of soil-borne chemicals and microorganisms through pinpoint leaks in distribution piping and valves (Moser 2000). In some cases, extremely low water system pressure can result in damage to pumps, including pump cavitation.

A study by Moser and the American Water Works Service Company (Moser 2000, Kirmeyer et al. 2001) attempted to determine the potential for intrusion of soilborne contaminants to the water main during episodes of low distribution system pressure. The first phase of the study was to survey water utilities to determine all of the following elements:

- Existence of an active leak detection program
- Ability to verify the amount of unaccounted-for water in the distribution system
- Estimate of percentage of water mains below the water table
- Percentage of meter boxes subject to periodic flooding

The second phase was a field study to determine the incidence and concentration of coliforms and pathogens in soil and water in the immediate vicinity of water mains (Kirmeyer et al. 2001). Typically there are provincial and/or state regulatory specifications for the minimum separation distance between water mains and sewer lines. The standards, if rigorously followed, help minimize the potential for contamination of municipal water mains from leaking wastewater collection systems. Leakage from wastewater systems contains various human and animal pathogens (Geldreich 1996).

Results demonstrated high fecal coliform concentrations, recovery of culturable viruses (enteroviruses, Norwalk and Hepatitis A virus), and common soil organisms such as Bacillus species and *Clostridium perfringens* in the immediate external vicinity of the water mains sampled (Moser 2000, Kirmeyer et al. 2001). Computer surge modeling was used to evaluate the potential for intrusion of external contaminants into the municipal water system during hydraulic transients (Moser 2000, Kirmeyer et al. 2001).

Nonutility Fire Hydrant Use

There are various community organizations or events, not related to water utility activities, that use fire hydrants and municipal water. These include

- Fire flow
- Filling of street-cleaning equipment
- Sewer-flushing activities
- Filling trucks used for watering municipal trees and flower beds
- Use by construction crews
- Traffic accidents that damage or destroy the hydrant
- Vandalism
- Illegal connections and theft of water
- Recreational use by urban youth, which may or may not be sanctioned by the water utility (Deb et al. 2000)

Utilities can exercise some degree of control by issuing permits and portable water meters to organizations permitted to draw water from the municipal supply by means of fire hydrants. However, this is not always possible. Pressure surges and main breaks can be caused by rapid opening and closing of hydrants. Therefore, nonutility operators that are permitted to open fire hydrants must be properly trained to maintain the integrity of distribution water and to prevent intrusion of contaminants. Often there is poor record keeping by other agencies, hampering follow-up investigations. A case in point is fire flow. Fire crews contact the utility if increased water pressure is needed for firefighting. However, there may be no further record available to the utility that describes the actual water flow and duration of flow from the discharge hydrant used. Cross-connections can be created when fire hydrants are used for filling water tanker trucks or other storage vessels, resulting in backsiphonage of truck or tank contents to the municipal system (Deb et al. 2000, PNWS–AWWA 1996).

Management and Training of Staff

Properly trained water utility staff is one of the most effective means to ensure provision of safe drinking water and protect public health (National Drinking Water Clearinghouse 1998). Many provincial and state regulatory agencies require certification of water plant operators to ensure that the individual meets a minimum requirement of knowledge and technical skill to manage, operate, and maintain the community water supply (Delvecchio 1998). Similar training requirements exist for public health agencies, medical providers, and regulatory agencies.

Poorly trained, or noncertified, individuals pose a serious risk for a water utility and the community at large. The 2000 Walkerton, Ont., Canada, *Escherichia coli* O157:H7 outbreak investigation and subsequent judicial inquiry demonstrated this concern. The utility manager had admitted deficiencies in knowledge about requirements for basic water treatment and distribution maintenance (i.e., importance of proper record keeping, significance of laboratory results, importance of maintaining distribution system disinfectant residuals, etc.) (O'Conner 2002). This individual was not certified, but rather was grandfathered in to the provincial certification program for water plant operators (O'Conner 2002). This, coupled with specific environmental events just before the outbreak, resulted in a community waterborne *E. coli* O157:H7 outbreak where 7 individuals died and more than 2,300 became ill (O'Conner 2002).

Workforce literacy is as a major issue for many industries, including water utilities (Marlow and Beaudet 1992). Water utility operations use increasingly sophisticated technology and processes. Utility staff require training in basic computer skills, as well as up-to-date training in laboratory and process operation skills including disinfection, corrosion control, source water protection, and distribution infrastructure maintenance. They also must have an understanding of the relationship that these processes have in the protection of community health (Xie, Cole, and Long 1999). Basic literacy and comprehension skills include the individual's ability to routinely analyze and interpret data, the ability to execute the proper action based on the interpretation processes and conclusions, and the ability to interact with the community (Marlow and Beaudet 1992). It has been demonstrated that individuals with marginal literacy skills have been able to function in a workplace provided that the technology and skill level required by the job tasks do not change or that the individual does not change jobs (Marlow and Beaudet 1992). When there are changes in tasks as a result of development of new technology or workplace requirements, this individual may become a source for errors or total disruption of the process (Marlow and Beaudet 1992).

Small water utilities typically lack the technical or financial resources available to larger utilities. This situation can create limited opportunities for professional development and upgrading for operators of small water systems. As a result, some small systems have difficulties in complying with changing regulatory requirements for safe drinking water (Xie, Cole, and Long 1999). Small systems may not have sufficient numbers of trained, or cross-trained, staff to handle unexpected, prolonged emergency situations on a 24/7 duration (Wise 1998).

The issues of training, opportunities for professional development, staff resources, and actual number of staff available also exist for public health agencies and public health laboratories.

The community public health system requires a strong infrastructure, with properly trained staff and functional capacity, to be able to respond to threats to the health of the community or emergency situations (CDC 2002; Talbot, Barry, and Fok 2001).

EVENTS THAT CAN CAUSE DISTRIBUTION SYSTEM FAILURE

Failure Due to Effects of Violent or Catastrophic Events in Nature

Events such as violent wind storms (e.g., hurricanes, tornadoes, cyclones), earthquakes, floods, subsidence, landslides and mudslides, major snowstorms or ice storms, severe thunderstorms, volcanic eruptions, and so on, are difficult to predict and virtually impossible to prevent. Historic records are used to predict the probability of future natural disasters, conduct infrastructure vulnerability analysis, and develop a community emergency response strategy.

Increasing the development and settlement of high-risk disaster areas, such as floodplains, earthquake faults, coastal hurricane areas, unstable hillsides, and so forth, has resulted in a proportional increase in numbers of victims and rebuilding costs when a disaster finally strikes (auf der Heide 1989). Similar parallels have been observed with increased development in areas close to human-made hazards such as hazardous waste landfills, airports, nuclear power plants, and so on (auf der Heide 1989). Often the processes of development result in removal of natural protective barriers to environmental threats, such as coastal sand dunes and deforestation of slopes (auf der Heide 1989, PAHO 2000a, b).

Table B.2 summarizes short-term effects from several major natural disasters. Table B.3 presents a matrix of community impacts caused by natural disasters. The complexity of the community disaster or emergency plan is in direct proportion to the magnitude of risk from the event. Table B.4 summarizes criteria used to define community risk from disasters.

Disaster and emergency planning involves multiple local, state/provincial, and federal agencies. Developing the community emergency plan involves several steps. They include

1. Vulnerability assessment of risks to community-critical infrastructure and known (historical) disasters in the area
2. Guidance from local, state/provincial, and federal standards and regulations to enforce community standards
3. Development of a communication plan and warning systems
4. Testing of the emergency plan to ensure coordinated response mechanisms
5. Development of a contingency plan for resources that may be temporarily unavailable during the emergency
6. Design, construction, and maintenance of critical community infrastructure to withstand the force of the disaster
7. Development of long-term measures to reduce the magnitude of a disaster (e.g., surface water diversion, flood walls, and earthquake-stable structures) (GVRD 2002; Clark and Deininger 2000, 2001; PAHO 1998).

Table B.2
Short-term effects of major natural disasters

Community effect	Earthquake	High winds, no flooding	Tidal waves, flash floods	Slow-onset floods	Landslides	Volcanoes/lahars*
Deaths	Many	Few	Many	Few	Many	Many
Severe injuries requiring extensive medical treatment	Many	Moderate	Few	Few	Few	Few
Immediate increased risk of communicable disease	Low to moderate†	Low†	Low†	Low†	Low†	Low†
Damage to health facilities	Severe (actual structure and equipment)	Severe	Severe (likely localized)	Severe (typically equipment more often than actual structure)	Severe (likely localized)	Severe (actual structure and equipment)
Major population evacuation	Rare in rural areas; more common heavily damaged urban areas	Rare in rural areas; more common heavily damaged urban areas	Rare in rural areas; more common heavily damaged urban areas	Common	Common	Common
Damage to drinking water systems (treatment plant, distribution system, wells)	Severe	Light	Severe	Light to moderate	Severe, localized	Severe
Damage‡ to source water supply	Yes	No	Yes	Yes	Yes (localized)	Yes
Damage to wastewater collection and treatment systems	Severe	Light	Severe	Light to moderate	Severe, localized	Severe

Source: Adapted from PAHO 1998.

* *Lahar* collectively refers to volcanic debris flows, mud flows, and hyperconcentrated stream flow. Lahars can occur during volcanic eruptions, after eruptions, or in the absence of a volcanic eruption.

† Potential for increased communicable disease incidence is a potential risk following all major natural disasters. The probability will increase because of lack of water for hygiene and health, deteriorating sanitation, overcrowding in emergency shelters, etc.

‡ Refers to contamination of source water supply (e.g., sediments, debris, dead and decaying wildlife, domestic animals, fish).

Table B.3
Matrix of community impacts caused by natural disasters

Impact areas	Community impact	Earthquake	Hurricane	Flood	Tidal wave/ flash flood	Volcanic eruption
Drinking water supply and wastewater treatment infrastructure	• Infrastructure damage	1	1	1	2 to 3	1
	• Ruptured water/sewer mains	1	2	2	1	1
	• Damage to water sources	1	2	2	3	1
	• Interrupted power supply	1	1	2	2	1
	• Intrusion of contaminants	1 to 2	1	1	1	1
	• Interrupted/failed transportation	1	1	1	2	1
	• Staff shortage	1	1 to 2	2	2 to 3	1
	• System overload	3	1	1	3	1
	• Shortage of equipment, parts, supplies	1	1	1	2	1
Solid waste handling	• Infrastructure damage	1	2	2	3	1
	• Interrupted/failed transportation	1	1	1	2	1
	• Equipment shortage	1	1	1	2	1
	• Staff shortage	1	1	1	2	1
	• Intrusion/introduction of contaminants	1	1	1	2	1
Private home sanitation	• Structural damage	1	1	1	1	1
	• Flooding of homes	3	1	1	1	3
	• Contaminated water and food	2	2	1	2	1
	• Disruption of water, power, heating, fuel, waste disposal	1	1	1	2	1
Vector control	• Increase in vector breeding sites	1	1	1	1	3
	• Increase in human/vector contact	1	1	1	2	1
	• Disruption of vector-borne disease control programs	1	1	1	1	1
Food handling and supply	• Infrastructure damage	1	1	2	3	1
	• Food spoilage	1	1	1	2	1
	• Interrupted/failed transportation	1	1	1	2	1
	• Interrupted power supply	1	1	1	3	1
	• Flooding of facilities	3	1	1	1	3
Communications	• Infrastructure damage	1	1 to 2	1 to 2	2 to 3	1
	• Interrupted power supply	1	1	2	2	1
	• Interrupted/failed transportation	1	1	1	2	1
	• Disrupted/lack of phone, radio, TV, other communications in community	1	1	2	2	1

Source: PAHO 1998.
1 = Severe potential effects.
2 = Less severe potential effects.
3 = Least or no potential effects.

Table B.4
Criteria to define community risk from the occurrence of a disaster

Odds ratio	1:1,000,000 to 1:100,000,000 Minimal acceptable risk (severe event)	1:100 to 1:999,999 Reducible risk (major event)	<1:100 Unacceptable risk (catastrophic event)
Deaths	1 to 10	11 to 50	>50
Injuries	10 to 100	101 to 500	>500
Loss of earnings	1 to 30 days	31 to 90 days	>90 days
Environmental impact	Reversible within 5 years	Reversible after 5 years	Irreversible
Material damage	US$1 billion to US$5 billion	US$5 billion to US$250 billion	>US$250 billion

Source: Adapted from PAHO 1998.

Medical facilities often suffer serious structural damage during a natural disaster, jeopardizing the treatment of mass casualties. Even if there is little, or no, structural damage, the facility may be without power and water, or may suffer serious damage to medical equipment and staff (PAHO 2000b). In the absence of adequate preemergency planning, community medical facilities may find themselves to be the victim of a functional collapse during the actual emergency.

The incidence of infectious disease directly related to a disaster can be difficult to ascertain unless the community has a well-established infectious disease surveillance system (Bissell 1983). Medical and public health agencies typically become overwhelmed with short-term casualties directly caused by a disaster (Wasley 1995). Disaster epidemiology programs may only be able to focus on the short, postdisaster interval; however, given the incubation period of disease and potential development of an epidemic, the full impact of the disaster may not be realized for several months (Bissell 1983). When hospitals become overwhelmed, they may determine that providing direct medical treatment during and after the disaster is more important than reporting notifiable disease to the health department, and therefore change their resource allocations accordingly (Ogden et al. 2001).

Earthquakes

Water utilities located in seismic risk areas are vulnerable to disruption and failure of water treatment processes and the distribution system as a result of an earthquake. Other damage caused by earthquakes includes

- Creation of faults in rocks and subsoil
- Settlement of the ground surface
- Landslides and mudslides
- Damage to buildings including industrial facilities, shopping centers, hospitals and medical clinics, apartments, houses, etc.

- Damage to elevated tanks (i.e., water, fuel, and others)
- Damage to buried structures (i.e., tanks; water, sewer, gas, and fuel pipelines; tunnels; wells, etc.)
- Damage to dams and reservoirs
- Damage to elevated transmission towers and power lines
- Damage to roads, rail lines, and airports (PAHO 1998, PAHO 2000a, b)

The amount of damage inflicted by an earthquake is a function of the magnitude of the earthquake, measured by the Richter scale, the type of materials and construction methods used in the community, time of day of the earthquake, and the population density in the zone of influence (PAHO 2000a, b). The effects of earthquakes, as experienced by individuals and community structures, are typically described using the Modified Mercalli Intensity Scale (Louie 1999). Table B.5 compares the Richter scale and the Modified Mercalli Intensity Scale. Table B.6 summarizes examples of water distribution system failure due to earthquake damage.

Previous experience with the devastating effects of earthquakes can be successfully used in future planning for emergencies (PAHO 1998, McReynolds and Simmons 1995, Tanaka 1995). Shared experiences, and collaborative regional emergency response and mutual aid plans, ensure that regional utilities and agencies can act quickly in the event of a major natural disaster or emergency (Riordan 1995).

Violent Winds and Floods

Severe thunderstorms and rainfall, causing flooding in addition to wind damage, typically accompany violent windstorms such as hurricanes and tornadoes. Common effects on community structures by hurricanes and tornadoes include

- Partial, to total, destruction of buildings (e.g., broken windows, roof damage, chimney damage) due to extreme wind force
- Ruptures in pipelines in exposed crossings over rivers and streams due to strong currents
- Pipe rupture and damage due to landslides or mudslides and/or strong currents
- Flooding of tanks and underground pipes
- Underground pipe rupture due to earth subsidence caused by heavy rainfall and/or flooding
- Soil erosion
- Damaged power and communications structures and systems causing interrupted services and operations (PAHO 1998, auf der Heide 1989, NIH 1996)

Flooding may accompany the hurricane or tornado, or simply be the result of extremely heavy rainfall. Floods are responsible for approximately 40 percent of natural disasters around the world (Ogden et al. 2001). Flooding has been linked with community waterborne disease outbreaks, primarily due to contamination of surface water and groundwater sources, sewer backup, pipe rupture due to soil erosion, and contamination of buried tanks and chambers (Ogden et al. 2001, PAHO 1998, Kramer et al. 1996, Reid 1994, Geldreich 1996). Table B.7 compares the various windstorm warning scales.

Table B.5
Comparison between Modified Mercalli Intensity Scale for earthquake effects and Richter Scale for earthquake intensity

Scale			
Mercalli*	Richter†	Earthquake severity	Description
I	0 to 4.3	Mild	Movement recorded by instruments. People do not feel any earth movement.
II			People at rest or on upper floors of tall buildings may feel movement.
III			Shaking felt indoors. People outdoors may not realize earthquake is happening. Hanging objects swing back and forth.
IV	4.3 to 4.8	Moderate	Dishes, doors, windows rattle. Parked cars rock. Most people indoors feel movement. Some people outdoors feel movement. Earthquake feels like a heavy truck hitting a wall.
V			Most people feel movement. People awakened from sleep. Doors swing open and closed. Liquids spill from glasses. Pictures on wall move. Small objects move or are turned over. Trees may shake.
VI	4.8 to 6.2	Intermediate	Everyone feels movement. People walk unsteadily. Objects fall from shelves. Pictures fall from walls. Furniture moves. Wall plaster may crack. Trees and bushes shake. Damage slight in poorly constructed buildings. No structural damage.
VII			People have difficulty standing. Plaster, bricks, and tiles fall from buildings. Drivers feel cars shake. Some furniture breaks. Damage slight in moderate to well-built buildings. Considerable damage in poorly constructed buildings.
VIII	6.2 to 7.3	Severe	Car steering affected. Chimneys/towers may twist and fall. Tree branches break. Houses not bolted down may shift on foundation. Hillside may crack if ground is wet. Water levels in wells may change. Well-built buildings suffer slight damage. Poorly built buildings suffer severe damage.
IX			General panic. Damage to foundations. Houses not bolted down move off foundations. Well-built buildings suffer considerable damage. Some underground pipes break. Ground cracks. Reservoirs suffer serious damage.
X			Most buildings and foundations destroyed. Some bridges destroyed. Dams seriously damaged. Large landslides occur. Water thrown on the banks of canals/rivers/lakes. Ground cracks in large areas. Railroad tracks begin to bend.
XI	7.3 to 8.9	Catastrophic	Railroad tracks bend. Roads break up. Large cracks appear in ground. Rocks fall. Most buildings collapse. Some bridges destroyed. Underground pipelines destroyed.
XII			Total destruction. Ground moves in waves. River courses altered. Objects thrown in air. Large amounts of rock may move.

Source: Anon. 2002, Louie 1999.

* The Mercalli scale is a semiquantitative linear scale.

† The Richter scale is a quantitative logarithmic scale.

Table B.6
Examples of distribution system failure and consequences due to earthquake damage

Date	Name and location of earthquake	Richter (Mercalli) earthquake magnitude	Municipal damage	Deaths	Injuries	Reference
Sept. 1, 1923	Great Kanto earthquake; Tokyo and Yokohama, Japan	NA*	Widespread municipal fires. Buildings destroyed because of earthquake and fire. Tsunami and landslides followed earthquake. More than 694,000 homes destroyed. Dams, tunnels, sewers, canals, retaining walls, lighthouses, water mains, power systems, roads, railway tracks, telephone and telegraph systems, newspaper offices destroyed.	143,000	ND†	Cameron and James 2002
May 22, 1960	1960 Chile earthquake; southern coast of Chile	9.5 (NA)	More than 2,000,000 homeless. One-half billion U.S. dollars in damages. Widespread municipal damage to buildings, underground utility lines. Earthquake caused revival of 6 dormant volcanoes and created 3 new volcanoes. Earthquake caused Pacific-wide tsunami that caused 61 deaths and US$75 million damage in Hawaii, 138 deaths and US$50 million in Japan, US$500,000 damage along west coast of United States.	>2,000	3,000	U.S. Geological Survey 2002
Oct. 1, 1969	Santa Rosa earthquake; Santa Rosa, Calif.	5.7 (NA)	Minor damage to storage tanks, pumping stations, dams. Major damage to distribution system. Structural damage to homes and buildings	0	ND	Cloud et al. 2002
Feb. 9, 1971	Sylmar earthquake; San Fernando, Calif.	6.6 (VIII–X)	Major damage to hydraulic structures. Damage to dams, reservoirs, water tanks, main tanks, water pipes, sewers. Intake on Van Norman Lake dam destroyed. Greater than US$5 million property damage. Significant structural damage to buildings. Hospital collapses.	65	ND	SCEDC 2002
Dec. 23, 1972	Managua earthquake; Managua, Nicaragua	6.25 (V–IX)	Approximately 100 main breaks. Roofs of pumping stations collapsed. Storage tanks damaged. Loss of water service in part of city. Severe damage to buildings, homes, roads. More than 300,000 homeless.	10,000 to 12,000	ND	PAHO 1998
Sept. 19, 1985	Mexico City earthquake; Mexico City, Mexico	8.1 (VIII–IX)	Major damage to water distribution system, communication, power. Major damage to buildings and homes. Damage ~US$5 billion.	10,000	50,000	John Martin & Associates 2002
Sept. 10, 1986	San Salvador earthquake; San Salvador, Central America	5.4 (NA)	Approximately 2,400 main breaks (or 20% of water distribution system damaged).	1,400	ND	PAHO 1998

(continued)

Table B.6 (Continued)

Date	Name and location of earthquake	Richter (Mercalli) earthquake magnitude	Municipal damage	Deaths	Injuries	Reference
Oct. 17, 1989	Loma Prieta earthquake; San Francisco Bay area, Calif.	7.1 (VI–VII)	More than 12,000 persons displaced. More than 18,000 homes damage; 963 homes destroyed. 2,500 buildings damaged; 147 buildings destroyed. Collapse of overpass of Interstate 880 in Oakland. Collapse of section of roadbed of roadway leading to Bay Bridge. Localized power failures and water main ruptures (~150 main breaks). Damage at Rinconada water treatment plant, Lexington water reservoir. Estimated damage US$6.8 billion.	62	3,700	EQE Engineering Inc. 1989
Apr. 22, 1991	Limon earthquake; Limon, Costa Rica	7.4 (VI–VII)	Contamination of surface water by mudslides. Significant distribution system damage (at pipe joints), intermediate segments in body of pipe. Transportation system (rail/road) devastated. Bridge collapse. Significant building damage.	53	198	PAHO 1998, Youd 1993
Jan. 17, 1994	Northridge earthquake; San Fernando Valley, Calif.	6.8 (NA)	Extensive damage to underground utilities (gas, water, sewer), toppled high-voltage power lines. Water treatment plants damaged, resulting in temporary shutdown. Damage to roads, freeway overpasses. Damage to aqueduct no. 1, and aqueduct no. 2. Ruptured transmission mains. Damaged water storage tanks. Damaged commercial and residential buildings. Estimated damage US$15 billion.	57	1,500	Aurelius 1994
Jan. 17, 1995	Hyogo-ken Nanbu earthquake; Kobe, Japan	7.2 (IX–X)	103,521 buildings destroyed. 300,000 homeless. Bridges collapsed. Widespread ground failure. 70% of Kobe water system inoperable (>2,000 distribution and transmission system breaks). Surface water intakes damaged leaving 1.5 million people without water. Failure of pump station and treatment plant. Electrical system seriously disrupted. Gas system severely affected (>1,400 breaks). Cost of restoring basic functions approximately US$1 billion.	5,096	26,797	Somerville 1995
Sept. 30, 1999	Oaxaca earthquake; Oaxaca state, Mexico	7.5 (NA)	Extensive municipal damage. Water and sewer treatment plants damage in 542 cities in Oaxaca state.	7	ND	Disaster Relief 1999

* NA = Information not available from the reference cited.

† ND = Number of injuries not determined or no data available that specified the number of injuries.

Table B.7
Comparison of windstorm measurement scales

Wind speed (mph)	Beaufort scale description*	Category of hurricane†	Beaufort scale*	Saffir–Simpson hurricane scale†	Fujita tornado scale‡	Effects
0 to 1	Calm		0	Depression		Land: Smoke rises vertically. Water: Like a mirror.
1 to 3	Light air		1			Land: Rising smoke drifts; wind vane not moving. Water: Small ripples.
4 to 7	Light breeze		2			Land: Leaves rustle; can feel wind on your face; wind vanes begin to move. Water: Small wavelets; wind fills sail.
8 to 12	Gentle breeze		3			Land: Light flags extend; leaves and small twigs move. Water: Large wavelets; sailboats heel.
13 to 18	Moderate breeze		4			Land: Moves thin branches; raises dust, leaves, and paper. Water: Sailboats at full speed.
19 to 24	Fresh breeze		5			Land: Small trees sway. Water: Numerous whitecaps.
25 to 31	Strong breeze		6			Land: Large tree branches move; umbrella difficult to keep under control. Water: Whitecaps everywhere; large waves.
32 to 38	Moderate gale		7			Land: Large trees begin to sway; becomes difficult to walk. Water: Much bigger waves, some foam.
<35 39 to 46	Fresh gale		8	Tropical storm	Gale tornado (F0) (40–72 mph)	Land: Small branches broken from trees; walking difficult. Some damage to chimneys, TV antennas, roof shingles. Water: Foam in well-marked streaks; larger waves; edges of crests break off.

(continued)

Table B.7 (Continued)

Wind speed (mph)	Beaufort scale description*	Category of hurricane†	Beaufort scale*	Saffir–Simpson hurricane scale†	Fujita tornado scale‡	Effects
47 to 54	Strong gale		9			Land: Slight damage occurs to buildings; shingles blown off roofs. Water: High waves, dense spray; visibility affected.
55 to 63	Whole gale		10			Land: Large trees uprooted; considerable building damage. Water: Very high waves; heavy sea roll; surface white with spray and foam; visibility impaired.
64 to 74	Storm		11		Moderate tornado (F1) (73–112 mph)	Land: Extensive, widespread damage. Cars overturned. Carports destroyed. Trees uprooted. Peels surface off roofs. Moving cars can be pushed off roads. Water: Exceptionally high waves, small to medium ships obscured, visibility poor.
74+ (74 to 95)	Hurricane	1	12	Hurricane		Land: Moderate to extreme destruction. Primarily to mobile homes, shrubs, and trees. Some damage to poorly constructed signs. Some coastal road flooding and minor pier damage. Water: Waves 40+ ft (12+ m), air filled with foam and spray, visibility restricted. Surge 4–5 ft (1.2–1.5 m).
96 to 110		2				Land: Moderate damage. Some roof, door, and window damage of buildings. Considerable damage to shrubs and trees. Considerable damage to mobile homes, poorly constructed signs, and piers. Coastal and low-lying escape routes flood 2 to 4 hours before arrival of hurricane center. Water: Surge 6–8 ft (1.8–2.4 m).

(continued)

Table B.7 (Continued)

Wind speed (mph)	Beaufort scale description*	Category of hurricane†	Beaufort scale*	Saffir–Simpson hurricane scale†	Fujita tornado scale‡	Effects
111 to 130		3			Significant tornado (F2) (113–157 mph)	Land: Extensive damage. Some structural damage to small residences and utility buildings. Roofs torn off frame buildings. Damage to shrubs and trees (foliage blown off, large trees uprooted. Mobile homes and poorly constructed signs overturned or destroyed. Boxcars pushed over. Low-lying escape routes cut off by rising water 3 to 5 hours before arrival of hurricane center. Flooding near coast destroys smaller structures. Land lower than 5 ft (1.5 m) above sea level may be flooded inland 8 mi (12.9 km) or more. Evacuation of low-lying residences within several blocks of shoreline required. Water: Surge 9–12 ft (2.7–3.7 m)
131 to 155		4				Land: Extreme damage. Complete roof structure failures on small residences. Shrubs, trees, and all signs blown down. Complete destruction of mobile homes. Extensive damage to doors and windows. Low-lying escape routes cut off by water 3 to 5 hours before arrival of hurricane center. Major damage to lower floors of structures near the shore. Terrain lower than 10 ft (3 m) above sea level may be flooded, resulting in massive evacuation of residential areas up to 6 mi (9.7 km) inland. Water: Surge 13–18 ft (4.0–5.5 m).
>155		5			Severe tornado (F3) (158–206 mph)	Land: Catastrophic damage. Complete roof failure of many residences and industrial buildings. Some complete building failures. Metal buildings collapse from severe damage. All shrubs, trees, and signs blown down. Complete destruction of mobile homes. Severe and extensive window and door damage. Trains overturned. Forests and farmlands flattened. Heavy cars lifted off the ground and thrown. Major damage to lower floors of all structures less than 15 ft (4.6 m) above sea level and within 500 yd (457 m) of shoreline. Massive evacuation of residential areas on low ground within 5 to 10 mi (8 to 16 km) of shoreline. Water: Surge >18 ft (5.5 m).

(continued)

Table B.7 (Continued)

Wind speed (mph)	Beaufort scale description*	Category of hurricane†	Beaufort scale*	Saffir–Simpson hurricane scale†	Fujita tornado scale‡	Effects
					Devastating tornado (F4) (207–260 mph)	Land: Devastating damage. Few walls, if any remain standing, in well-built buildings. Structures with weak foundations blown off for some distance. Large steel and concrete missiles thrown far distances. Cars thrown.
					Incredible tornado (F5) (261– 318 mph)	Land: Incredible damage. Homes leveled with all debris removed. Well-built structures lifted off foundations and carried considerable distances to disintegrate. Large infrastructures (e.g., schools, motels) have significant exterior damage with walls and roofs gone. Automobile-sized missiles fly through air in excess of 109 yd (100 m). Trees debarked.
					Inconceivable tornado (F6) (319–379 mph)	This wind speed unlikely. The small area of damage produced will likely not be distinguished from F4 and F5 winds surrounding F6 tornadoes. Damage would be catastrophic and complete.

Source: Adapted from NOAA 2002a, b; SPC 2002.

* The Beaufort scale is a wind-force scale first created in 1838 and subsequently revised to include wind velocities. It is commonly used by vessels at sea (NOAA 2002a).

† The Saffir–Simpson scale was created in 1969 and is used to estimate potential property damage and flooding from hurricanes (NOAA 2002b).

‡ The Fujita tornado scale was developed in 1971 and is used to classify tornadoes according to wind speed and damage caused (SPC 2002).

Volcanic Activity

Volcanoes occur throughout the world. Many have significant numbers of people living in close proximity. Although volcanic soil is extremely fertile, enticing agricultural and community development, the risk of eruption remains (PAHO 2000a). During a volcanic eruption, a chain of events can occur, including

- Eruption of ash, dust, gases, rocks, and lava
- Flooding
- Snowfall, mudslides, or landslides
- Additional seismic activity

Water utilities in the immediate vicinity of active volcanoes can be subject to contamination of surface water supplies due to volcanic eruption. Often the eruption results in massive deposits of ash, as well as destruction of vegetation and wild and/or domestic animals, which may accumulate and decompose in source water supplies (PAHO 2000a). Heavy ashfall can block transportation routes, as well as plug machinery, filters, pipelines, water treatment structures, and wastewater treatment facilities (Tilling, Topinka, and Swanson 1990). Areas downwind of the erupting volcano may experience similar disruptions. Buried structures, including distribution pipes, can be seriously damaged or crushed by large, fast-moving lava flows.

Failure Due to Deliberate Human-Caused Disruptions, Sabotage, Terrorism, or Acts of War

As the millennium (year 2000, or Y2K) approached, there were massive efforts to ensure that computer systems throughout the world were Y2K-ready (Ladner and Leake 1999). As part of Y2K contingency planning, public and private critical-infrastructure agencies and governments developed extensive preparedness and response strategies to minimize the potential for community disruption from Y2K failure (Koplan 2002, FEMA 2000, DoD 1999).

Recent events of terrorism and world instability have underscored the vulnerability of water utilities to deliberate acts of sabotage, terrorism, or acts of war (Clark and Deininger 2001, Luthy 2001). Disruption of source water supplies and/or treated water systems has been used for hundreds of years as a means of severely disrupting a community, as well as for increasing the potential for military conquest (Gleick 2000, Chalecki 2001). Communities rely on an adequate, safe, and constant supply of drinking water to survive.

Water utilities throughout the developed world share common areas of vulnerability:

- Increasing use and storage of hazardous materials (i.e., chlorine, caustic soda, and others) in centralized facilities at the water treatment plant
- Increasing reliance on computer technology to control critical systems at the water treatment facility and throughout the distribution system
- Dependence on a reliable source of power and fuel to run the water treatment plant and distribution system pumps
- Aging distribution infrastructure systems (Lancaster-Brooks 2002, Denileon 2001, Lacombe and Keyes 2000)

Knowledge of common vulnerabilities can be used for the design of terrorist strategies, as well as utility protection programs. On Jan. 24, 2001, the U.S. Federal Bureau of Investigation (FBI) issued a warning that a signed threat had been received from a North Africa–based terrorist group indicating that they were planning to disrupt water operations in 28 American cities (Denileon 2001). More recently, the FBI issued a cyber terror warning that members of al-Qaeda were seeking information on the Internet about water plant supervisory control and data acquisition (SCADA) systems (Poulsen 2002).

Activities or actions that can disrupt water supply are debilitating to the community that is under attack (Denileon 2001). Disruptions to water supply can be the result of vandalism, sabotage, terrorism, or acts of war. Sites for supply disruption can include open-water reservoirs or impoundments, water storage facilities, water treatment plants, source water intakes, and distribution system components (Clark and Deininger 2001, Denileon 2001, Lancaster-Brooks 2002). Refer to Table B.8 for examples of military disruptions of community water supplies.

Motivation for seeking deliberate water supply failure or destruction is not always clear. Vandalism has been an ongoing problem for utilities and can range from acts such as placing graffiti on utility or distribution system structures to damaging water meters and forced entry into buildings, storage reservoirs, and so forth (AWWA 1994). Teenage or young adult pranksters (AWWA 1994) have traditionally been linked with vandalism.

Deliberate sabotage and/or terrorism efforts can be the result of any of the following:

- Disgruntled current or former employees
- Contractors seeking revenge
- Disenfranchised individuals that may target victims for their ethnicity, beliefs, or for unknown reasons
- Radical ideologists wanting to make a statement
- Religious fanaticism
- Nationalist fanaticism
- Third-party-sponsored fanatics (i.e., sponsored by states such as Libya and others) (Hebert 2001, Denileon 2001)

Deliberate sabotage/terrorism efforts are typically the work of organized, trained, funded individuals and groups rather than pranksters, although "copy-cat" efforts by unaffiliated persons have been reported.

COMMUNITY CHARACTERISTICS THAT CREATE VULNERABILITY TO WATERBORNE CHEMICAL AND BIOLOGICAL AGENTS

Risk and Vulnerability

Definitions of Risk

It is the responsibility of public health agencies to reduce and eliminate risks of disease and injury, and promote health (Talbot, Barry, and Fok 2001). The term *risk* is used to express the probability for an individual to develop an adverse effect as the result of an exposure or an activity (Cohrssen and Covello 1989; Foster, Bernstein, and Huber 1993; Last 1987). The term *risk factor* can mean any or all of the following:

Table B.8
Examples of military disruptions of community water supply

Year	Countries involved	Reason for disruption	Description of disruption
1503	Italy (cities of Florence and Pisa)	Military conflict	Planned diversion of Arno River away from Pisa, removing potential source water supply
1924	United States (Owens Valley, Los Angeles, Calif.)	Control of water resources using terrorism	Repeated bombing of Los Angeles aqueduct to prevent diversion of water from the Owens Valley
1940 to 1945	Multiple countries	Military conflict and target	Repeated bombing of hydroelectric dams during World War II causing flooding, disruption of power supplies and drinking water supplies
1948	Palestine, Israel	Military conflict	Arab forces cut off West Jerusalem's water supply in first Arab–Israeli war
1950s	Korea, United States, United Nations	Military conflict and target	Repeated bombing of dams causing flooding, and disruption of power and water supplies
1960s	North Vietnam, United States	Military conflict and target	Bombing of irrigation water supply systems causing flooding
1980 to 1988	Iran, Iraq	Military tool	Iran diverts surface water to flood Iraqi defense positions
1981	Iran, Iraq	Military conflict and target	Iran bombs hydroelectric facility in Kurdistan causing widespread power outages in Iraq
1982	Israel, Lebanon, Syria	Military conflict and tool	Israel cuts off water supply to Beirut.
1988	Angola, South Africa, Cuba	Military conflict and target	Cuban and Angolan forces bomb Calueque Dam causing power disruption. Water transmission line to Owamboland is destroyed.
1991	Iraq, Kuwait, United States	Military conflict and target	Iraq destroys most of Kuwait's desalination capacity during retreat in Gulf War.
1991	Iraq, Kuwait, United States	Military conflict and target	Allied coalition bombs Baghdad's water supply and wastewater treatment systems.

(continued)

Table B.8 (Continued)

Year	Countries involved	Reason for disruption	Description of disruption
1992	Bosnia (Bosnian Serbs)	Military conflict and target	During Serbian siege of Sarajevo, Bosnia, and Herzegovina, all electrical power and water supplying Sarajevo cut off. Bosnian Serbs take control of water valves regulating flow from wells, cutting off more than 80% of city's water.
1993–2002	Iraq	Military tool	Saddam Hussein allegedly orders poisoning and/or draining of water supplies of southern Shiite Muslims.*
1997	Singapore, Malaysia	Political tool	Malaysia threatens to cut off water supply to Singapore as a result of Singapore's criticisms of policy in Malaysia.
1999	Zambia (Lusaka)	Terrorism	Bomb blast destroys main water transmission line for city of Lusaka, disrupting water supply for 3 million people.
1999	Yugoslavia	Military conflict and target	North Atlantic Treaty Organization (NATO) targets water utilities and shuts down water supplies in Belgrade.
1999	Kosovo	Military conflict and political tool	Serbian engineers shut down water system in Pristina prior to NATO occupation.
1999	Angola	Terrorism	100 bodies are found in four drinking water wells in central Angola.
1999	East Timor	Terrorism	Militia opposing independence kill pro-independence supporters and throw bodies into drinking water wells.
1999	Kosovo	Terrorism	Serbs contaminate water supplies and wells by throwing bodies of Kosovar Albanians into local water wells.

Source: Adapted from Gleick 2000.
* Data from Elkhamri et al. 2005.

- An attribute or exposure that is associated with an increased probability of a specified outcome
- An attribute or exposure that increases the probability of a specified outcome
- A determinant that can be changed through intervention, reducing the probability of a specified outcome (Valanis 1992)

The term *risk assessment* refers to the processes used to quantify and evaluate a risk. Risk assessment includes all of the following:

- Identification of the hazard
- Quantification of the hazard
- Assessment of the relationships between dose of hazard and host response
- Assessment of exposure received and duration of exposure
- Estimation of risk potential in a given situation

Risk assessments are based on the evaluation of existing and historical data; therefore, the conclusions drawn are directly influenced by the quality of the data being examined. It is crucial in risk assessment to clearly define sources or error, and uncertainty or bias in the information being evaluated (Potter 1999, Kindred 1996). Failure to do so will result in inaccurate estimation of the nature and magnitude of the risk factor under investigation. Risk assessment ultimately results in strategies for risk management.

Risk management is the process of "weighing policy alternatives and selecting the most appropriate regulatory action, integrating the results of risk assessment with engineering data, and with social, economic, and political concerns to reach a decision" (NRC 1983, p. 3). Development of regulatory policy and action often requires a balancing action between target and competing risks within a rigorous, scientific, analytical framework (Putnam and Graham 1993). As an example, concern about long-term carcinogenic effects of exposure to chlorinated disinfection by-products must be carefully weighed against short-term effects of exposure to infectious water-borne microorganisms.

The phrase *risk perception* refers to individual or societal attitudes and responses to potential or defined risk factors (Cohrssen and Covello 1989, Scala 1991). Risk perception is strongly influenced by the state of knowledge about the risk factor, fear factors, ability for control of risk factors and exposure, equity in exposure to risk, control over exposure to risk, and probability of widespread catastrophic outcome (Scala 1991, Cohrssen and Covello 1989). Typically there are significant differences in risk perception between the general public and experts (Scala 1991).

Public reaction to risk is often defined, and refined, by the news media, especially when sensationalism affects reporting of the risk factor or potential for risk (Foster, Bernstein, and Huber 1993). This has created a phenomenon known as "phantom risk," which is a risk that is not yet proven or not able to be proven but creates the foundation for tremendous public anxiety and dread and often results in public litigation against sources of that fear (Foster, Bernstein, and Huber 1993, Whelan 1993). The layperson's perception of risk is often based on emotional rather than objective evidence (Glicker 1992). Public fears and apprehensions about the potential for personal or community risk can lead to public statements such as

- "We're being poisoned and killed by chemicals, radioactivity, and microorganisms."
- "The country's health has never been worse, and will continue to deteriorate."

- "Big business is responsible for all environmental problems and only cares about their short-term profits and their bottom-line."
- "Governments almost always side with big businesses at our expense."
- "Little people like us are always the victims of corporate greed."
- "Current trends in disease and death are the ultimate price we must pay for technology and the good life."
- "All chemicals are poisons. New chemicals are worse poisons. They must be highly regulated or stopped."
- "All microorganisms are bad and will make us very sick or kill us."
- "Someone, or something, is totally responsible for our current state of ill health and we deserve to be compensated by them." (Whelan 1993)

Statements such as these serve to justify and reinforce the original fear and increase, or create, new fear in other members of the community.

Measurement and Assessment of Risk

Processes that are used to conduct risk assessment involve determining the magnitude of the risk, potential for exposure to risk, and outcomes of that exposure over time. Water utilities, through mandated regulatory water quality monitoring programs, routinely quantify the magnitude of potential biological, chemical, and physical hazards in source water, plant effluent, and throughout the distribution system. Public health agencies, through notifiable disease monitoring programs and/or outbreak investigations, routinely quantify the exposure and resulting outcome from regulated hazards in the community. Epidemiological and toxicological studies attempt to associate specific risk factors and exposure to observed disease patterns (Bull and Bull 2001).

The fundamental principle of an epidemiological study design is to create a fair, unbiased contrast between a group exposed to a given risk factor and/or intervention with a group that is not exposed to the risk factor and/or intervention (Jekel, Elmore, and Katz 1996, Bull and Bull 2001). The first step in creating the study design is formulation of the hypothesis. If the test results and evaluation appear consistent with the hypothesis, the probability that the hypothesis is true is strengthened (Jekel, Elmore, and Katz 1996). Table B.9 summarizes relative strengths and weaknesses of common epidemiological study designs.

Surveillance

Quantification of waterborne hazards is a function of the best available analytical technology, sensitivity of the technology, and costs of the technology. Advances in technology for hazard identification and characterization typically proceed at a faster rate than regulatory standards for drinking water (Jones 1997). As standards and legislation change, expectations and definitions of acceptable water quality also change (Jones 1997). Technological advances result in revised or new requirements for potable water, which changes the standard of practice and regulation. Increased expectations drive the need to improve technology. This cycle of development is illustrated in Figure B.2.

Water utilities may have neither the financial/capital resources nor the expertise to routinely monitor for every single waterborne risk factor (Allen, Clancy, and Rice 2000). Regulatory requirements in North America dictate the minimum number of water samples, based on

Table B.9
Factors that may influence the occurrence and severity of disease in individuals

- Age at time of exposure to disease-causing agent
- Chronic disease (preexisting or coexisting)
- Physical status at time of infection or exposure
- Immune status of host at time of exposure or infection
- Immune response of host to disease-causing agent
- Mechanism of disease production
- Host socioeconomic factors
- Host psychosocial factors
- Alcoholism and/or drug use
- Perception of illness by patient
- Route of entry of disease-causing agent
- Dose and duration of exposure to disease-causing agent
- Host immunodeficiency status (e.g., natural, drug-induced, disease-induced)
- Pregnancy
- Trauma
- Stress

Source: Adapted from Evans 1993.

community size, and routine parameters that all water utilities must monitor. To meet these requirements, water utilities routinely conduct source water, in-process water, and distribution water surveillance for regulated parameters and indicator organisms. Results are reported to provincial/state environmental regulators.

Quantification of public health hazards is also a function of the best available technology, sensitivity, and cost of implementing technology for a health jurisdiction. Public health regions may lack sufficient resources or staff to routinely monitor for every single cause of disease in the community on an active basis.

Many health agencies routinely rely on established surveillance programs to monitor the health status of their community and, by extension, exposure to risk factors for disease (Thacker and Berkelman 1988; Talbot, Barry, and Fok 2001). There are four common systems of surveillance used:

1. Notifiable disease reporting
2. Hospital-based or physician-based surveillance
3. Public health laboratory-based surveillance
4. Population-based surveillance

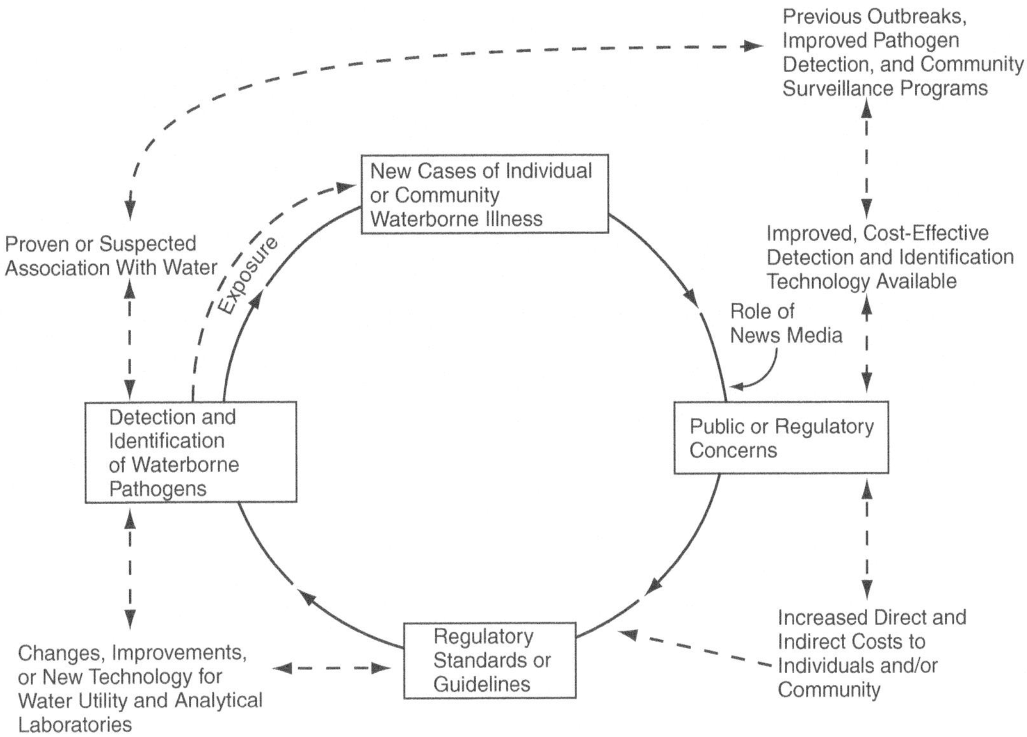

Source: Adapted from Jones 1997; Emde, Geldreich, and Talbot 2001.

Figure B.2 Cycle of drinking water issues and waterborne disease

All of these systems rely on community physicians and medical care facilities, the willingness or need for individuals to seek medical care, and the ability of the public health or clinical laboratory to isolate the causative agent from clinical or environmental specimens. Surveillance data can be used to:

- Detect epidemics of disease in the community
- Document the distribution and spread of a health event
- Provide quantitative estimates of the magnitude of a health event or problem
- Test the hypothesis
- Evaluate control and prevention measures
- Plan for the future (Talbot, Barry, and Fok 2001)

Outbreak Definition and Outbreak Investigation

An outbreak is defined as a particular point in time and geographic location where at least two individuals have the same disease, and/or the same clinical symptoms, and/or the same pathogen (IAMFES 1996). Outbreak investigation has ten key components, which include

1. Development of a clear, easily understood case definition
2. Confirmation of cases that meet the case definition
3. Establishment of community endemic rates for the disease in question

4. Estimation of the descriptive epidemiology of the cases
5. Generation of a hypothesis about the source, route of exposure, and incidence and prevalence of disease in the community
6. Testing of the hypothesis
7. Collection and analysis of environmental samples from suspected source of disease and clinical specimens from case-control groups
8. Initiation of environmental and public health control measures to limit the outbreak
9. Communication with the public and the media
10. Evaluation of the findings to create new detection and prevention strategies (Tillett, de Louvois, and Wall 1998; IAMFES 1996; Thacker and Berkelman 1988)

Outbreak investigation is different from routine epidemiological surveillance in that during an outbreak investigation the purpose of the initial investigation is to quickly determine

- The problem
- The magnitude and sources of community exposure
- The time periods of exposure
- The locations of exposure
- Population groups involved
- Confirmation of reported cases
- Establishment and implementation of any temporary measures needed to limit further spread of the outbreak (Thacker and Berkelman 1988, Reingold 1998)

Characteristics of Vulnerable Groups in the Community

Vulnerability of Individuals

When individuals become ill, irrespective of the cause, their social status is immediately transformed from "healthy" to "sick." This change in status causes altered perceptions, obligations, and, in some cases, privileges for that individual and the community at large (Basch 1990). Disease and illness are not randomly distributed throughout the population. Rather, disease is a function of

- The natural environment of the individual and/or community
- Opportunities for contact with disease-causing agents
- Level and duration of exposure to disease-causing agents
- Family and community patterns of disease that are the result of genetic changes
- Age
- Underlying illness or medical conditions
- Availability of medical care and treatment (Basch 1990, Evans 1993, Cohrssen and Covello 1989)

Table B.10 summarizes factors that can influence the occurrence and severity of disease in individuals.

Table B.10
Examples of waterborne disease outbreaks related to distribution system infrastructure failure

Year and location	Type of failure	Cause	Contaminant	Community impact	Health impact	Reference
1970, Olivet, France	Insufficient distribution system disinfectant residual	Insufficient chlorine dosage; no contact with water reservoir	*Escherichia coli, Streptococcus faecalis, Clostridium perfringens*	Contamination of water supply; inactivation of waterborne pathogen	52 cases of infectious hepatitis	Gavan and Nutt 1970
1971, Pico Rivera, Calif.	Sewage contamination of municipal distribution system; chlorination equipment failure	Insufficient chlorine; sewage contamination of water main	Unspecified total and fecal coliforms	Not specified	824 cases of diarrhea, nausea, vomiting, cramps, fever, hematochezia	Mahoney et al. 1974
1980, Rome, Ga.	Cross-connection between textile plant and municipal distribution system	Unreliable municipal chlorination; periodic pressure drops in municipal distribution	Norwalk virus	Not stated	1,500 persons ill with gastrointestinal symptoms	Kaplan et al. 1982
1993, Gideon, Mo.	Cross-connection between contaminated industrial water tank and municipal distribution system; wellhead contamination from run-off	Cross-connection; uncoordinated municipal mains flushing	*Salmonella typhimurium*	Community hydrant flushing in response to taste and odor complaints	Approximately 600 ill (diarrhea); 7 deaths	Clark et al. 1996a, b

(continued)

Table B.10 (Continued)

Year and location	Type of failure	Cause	Contaminant	Community impact	Health impact	Reference
1995, Fife, Scotland	Sewage-contaminated stream water entry to municipal distribution system	Malfunctioning distribution pump	*E. coli* O157:H7, *Campylobacter jejuni, Campylobacter coli*	Contamination of village water supply; extensive flushing program	711 residents experienced gastrointestinal symptoms; 633 of ill residents met the outbreak case definition (and cases were confirmed in a laboratory); 2 of these cases developed sequelae (haemolytic uraemic syndrome)	Jones and Roworth 1996
1996, Uzbekistan	Lack of piped water; deterioration of water treatment plant; lack of chlorine residual; reduced municipal system water pressure; cross-connections; poor community sanitation	Distribution system insufficiency; no chlorine residual; low system pressure	Unspecified	Contamination of community water supply	Community gastrointestinal disease (diarrhea)	Semenza et al. 1998
1997, Rochester, N.Y.	Contaminated ice machine; contaminated hospital hot- and cold-water taps	Not determined	*Legionella pneumophila* serogroup 6	Hospital shock-chlorination of water and ice-making system	1 individual developed nosocomial pneumonia after 3 months in intensive care unit	Graman, Quinlan, and Rank 1997

Traditional population risk groups have included

- Infants and young children
- Elderly individuals
- Residents of medical care facilities (i.e., hospitals, nursing homes, long-term care facilities, etc.)
- Pregnant women
- Surgical and transplant patients
- Persons with chronic disease
- Persons who are immunocompromised, immunosuppressed, or immunodeficient (Levine et al. 1991; Chow et al. 1995; Bartlett et al. 1985; Gerba, Rose, and Haas 1996)

Risk groups are changing and increasing in modern society and can be found in larger numbers both within and outside of medical care facilities. Current trends toward shortened hospital stays and longer periods of home care increase the importance of home water systems, yet those systems increasingly become routes for dissemination of waterborne contaminants. This can result in the epidemiology of waterborne infection in the community becoming more difficult. Symptoms of waterborne infection can be masked by other health conditions or may not be severe enough to warrant a physician visit. Also, there may not be recognition of the link to the water distribution system; thus, the actual incidence of waterborne infection may tend to be underestimated. Figure B.3 illustrates this cycle.

Vulnerability of Institutions and Industries

Institutional facilities such as correctional facilities, boarding schools, nursing homes, hospitals, secondary and tertiary medical care facilities, military installations, religious boarding facilities (e.g., convents, seminaries), schools, day-care facilities, and others can be found in many communities throughout the world. These types of facilities are often used for epidemiological studies, as populations are easily identified and tracking of cases is simplified by the nature of institutional confinement. A recent field study examining the incidence of waterborne gastrointestinal disease in two North American cities used active surveillance at day-care facilities, elementary schools, correctional facilities, nursing homes, and convents to determine the incidence of community waterborne disease over a 12-month period (Emde et al. 2001). Although there were no discernable waterborne disease outbreaks in either city, the study surveillance program was able to quickly detect outbreaks of foodborne illness, a Norwalk virus, chickenpox, influenza and tonsillitis, and head lice. The time frame was such that there was an immediate outbreak investigation and interventions implemented to stop each outbreak (Emde et al. 2001).

Persons at institutional facilities enter from the larger environment where pathogens typically are present in much lower concentrations than that found within the institution (Garrett 1994). A significant percentage of individuals in many of these institutions are temporarily or permanently immunocompromised, increasing their susceptibility to disease agents. Structures and practices of the institution can serve to amplify very low levels of environmental pathogens (Garrett 1994). The general trend is that once water enters a building's plumbing system, microorganisms present in the municipal water supply will amplify within various components of the building's plumbing system. For example, *Mycobacterium avium*, one of the most common

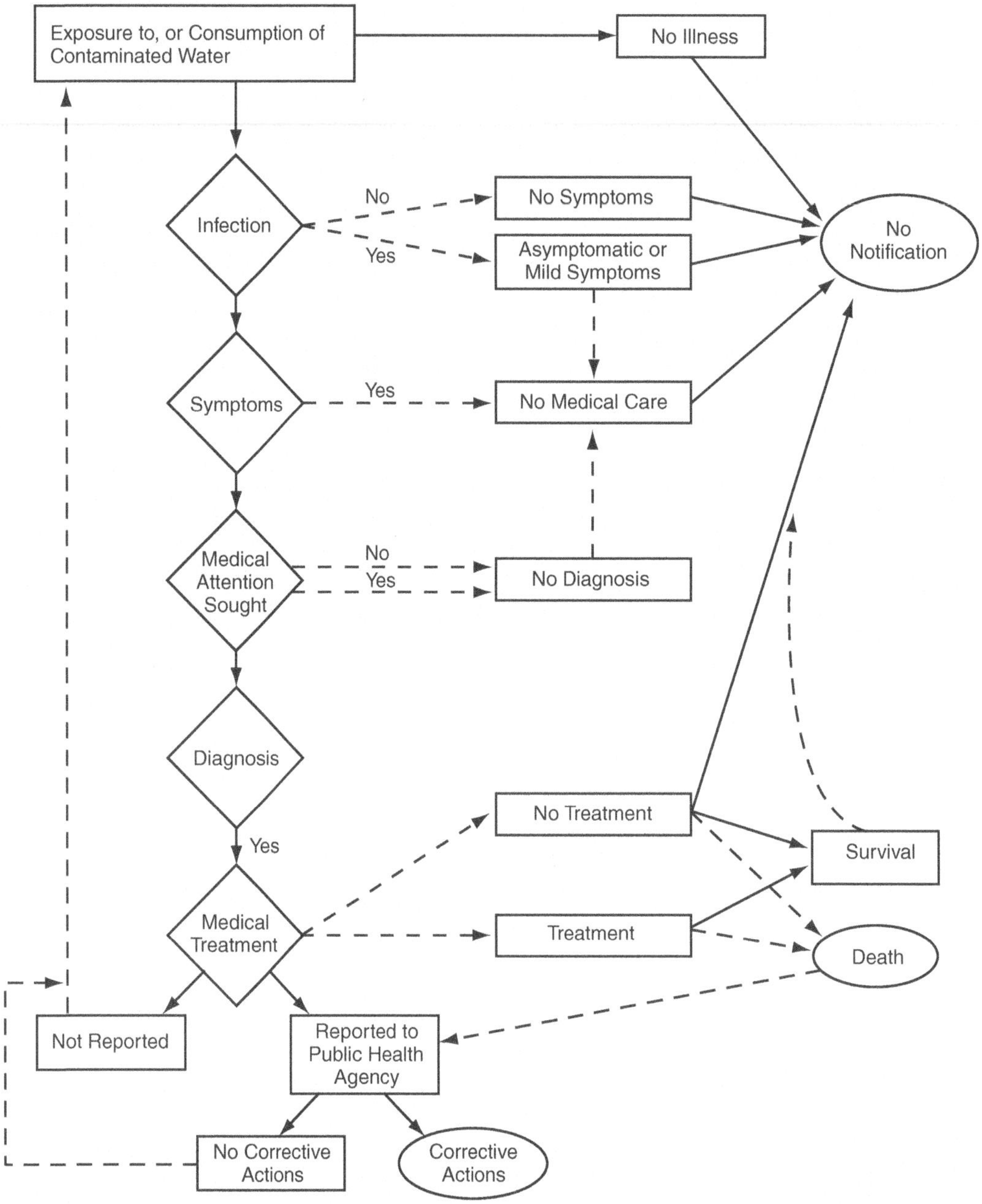

Source: Adapted from Emde et al. 2001.

Figure B.3 Cycle of events from exposure to waterborne contaminants and pathogens to public health notification

causes of pneumonia in patients with acquired immunodeficiency syndrome, has been shown to be concentrated by hospital hot-water systems with the hot water serving as a vehicle for transmission of this pathogen (du Moulin et al. 1988). Similarly, there are many reports of nosocomial legionellosis, where *L. pneumophila*, present at low levels in municipal water supplies, entered the hospital, amplified within the plumbing systems, and was subsequently able to cause infection (Blatt et al. 1993, Joly and Alary 1993, Breiman 1993). Ruggenini et al. (1989) determined that

24 percent of hospital patients diagnosed with pneumonia at autopsy actually had nosocomially acquired *L. pneumophila*, which entered the hospital from the municipal water supply.

Many industries use potable water, without further in-house treatment, as part of their manufacturing processes or as actual product. Bottled-water and ice manufacturers are especially vulnerable to events that contaminate municipal water supplies. Both of these products are manufactured according to regulations and specifications for food rather than for drinking water. Bottled water is typically the product of choice during community disasters or boil-water orders. In addition, bottled water is often perceived to be more pure or of a higher quality than municipal drinking water, and, as determined by a recent study, more than 50 percent of Americans use bottled water (Hamilton 1996, Hurd 1993). Ice manufactured during waterborne disease outbreaks has been used for retrospective confirmation of waterborne pathogens. There have been numerous outbreaks of waterborne disease from bottled water (Warburton 1993, Kramer et al. 1996, Blake et al. 1977).

Other industries, such as food manufacturers, are also vulnerable to changes in municipal water quality. Manufacturers of ready-to-eat, packaged vegetables use municipal water supplies to wash their product before packaging. Other manufacturers, such as the meat-packing, dairy, and beverage industries, will use municipal water to wash their products before further processing, to clean processing equipment, and for worker hygiene. Grocery stores use municipal water to spray vegetables, clean equipment, and for worker hygiene. The food service industry uses municipal water to prepare foods, clean equipment, to serve to customers, and for worker hygiene.

HAZARDS ASSOCIATED WITH DISTRIBUTION SYSTEM FAILURES

Biological Contaminants

Drinking water is not sterile, a fact recognized by the regulatory limits set for heterotrophic plate count (HPC) in drinking water—500 cfu/mL. Water utilities strive to produce potable water that is of minimal public health risk and of acceptable taste and aesthetic quality to the consumer. Routine utility monitoring for biological contaminants uses indicator organisms, as it is neither practical nor possible to determine the absolute concentration of all potential waterborne pathogens that may be present. Total and fecal coliforms, as well as HPC, have been used as indicators of choice (Geldreich 1996). However, these microbial indicators do not correlate absolutely with the presence of all potential waterborne pathogens, especially protozoans (Craun, Berger, and Calderon 1997).

Interpretation of routine microbial monitoring results assumes a homogeneous distribution of the indicator organism in the sample analyzed (Gale, van Dijk, and Stanfield 1997). As microorganism distribution in water is typically heterogeneous, it possible to miss aggregates of organisms, especially if low numbers of small-volume samples are analyzed (Gale 1996). This can create critical errors for the detection of waterborne pathogens.

Sources of Microorganisms in the Distribution System

Sources of distribution system biofilm organisms are varied. These organisms can arise from the original source water, enter during the treatment process, or reflect some form of post-treatment contamination event (Emde, Smith, and Stanley 1995). A portion of microorganisms present in untreated surface water may pass through treatment barriers as a result of one of the following circumstances:

- Short-circuiting of the treatment system
- Turbidity spikes
- Association with particulates or aggregates of organisms
- Unknown reasons

When these organisms enter low-nutrient environments, such as potable water, there are several possible fates. The organisms can adapt to their new environment, they can enter into some state of dormancy, or they can die (Roszak and Colwell 1987). This becomes a problem not only for enumeration technology, but also in understanding how organisms adapt to and survive in nutrient-limited environments. Failure to understand these processes can cause serious underestimation of the numbers and types of organisms present in the environment being sampled. *Standard Methods for the Examination of Water and Wastewater* (APHA, AWWA, and WEF 2005) acknowledges the limitations of best available analytical methods.

Microorganisms tend to adapt quickly to changes in their environment by a variety of genotypic and phenotypic strategies (Roszak and Colwell 1987). Organisms in aquatic environments are subjected to changes in temperature, substrate availability, pH, metals concentration, oxygen levels, pressure differences, and presence of other competing or predatory microorganisms and higher organisms (Olson 1982).

Organisms in an oligotrophic environment, such as potable water mains, tend to attach to solid surfaces, such as particulates, pipe or valve walls, and so forth (Amann et al. 1996, Block et al. 1994, Costerton et al. 1987, Stevenson 1978). Concentrations of some critical nutrients at the solid–liquid or solid–air interface are considerably higher than in the bulk phase (Costerton et al. 1987, Stevenson 1978). The organism attaches by means of extracellular polysaccharides (glycocalyx), or because of interactions between microbial surface charges with charges on the particulate substratum (Costerton and Boivin 1991). The adhesion may be reversible or irreversible (Sutherland 1989). Waterborne pathogens have been demonstrated to survive and persist in water distribution system biofilms (Camper et al. 1998, Kirmeyer et al. 2001, Geldreich and LeChevallier 1999, Geldreich 1996, Chang et al. 1960).

Distribution system biofilms generally are not uniform over the entire interior surface of the pipe. Microorganisms tend to preferentially attach to other attached organisms, or polymers present on the solid support (Characklis and Cooksey 1983). Parameters such as hydrophobicity, surface topography, electrostatic charge, hydraulic shear at the liquid–solid interface, and surface free energy will determine the numbers and types of organisms able to attach at any given location in the pipe (Characklis and Cooksey 1983).

Waterborne Disease Outbreaks

Both the United States and Great Britain have extensive waterborne disease monitoring programs, which has resulted in improved detection of waterborne outbreaks and characterization of risk factors involved in the outbreak. The first comprehensive framework for waterborne outbreak detection and evaluation was established in the United States in the 1930s (Gorman and Wolman 1939). The incidence of waterborne disease was found to be extremely low (Kramer et al. 2001). The authors of this study felt that the survey results indicated extreme underreporting rather than extremely low rates of waterborne disease (Kramer et al. 2001).

The U.S. waterborne disease monitoring program is capable of differentiating root causes of reported waterborne outbreaks and, as a result, can identify distribution system deficiencies as a risk factor (Craun and Calderon 2001). During the period from 1971 to 1998, approximately 30 percent of U.S. outbreaks were directly related to distribution system deficiencies (Craun and Calderon 2001). Chemical and microbial contaminants entering the distribution system because of main breaks, cross-connections, and backsiphonage caused approximately 51 percent of the outbreaks in municipal water systems (Craun and Calderon 2001). Table B.1 summarizes incidences of chemical and biological intrusion into the distribution system due to backsiphonage and cross-connections. In the majority of cases presented in Table B.1, a municipal main break was the pivotal event for contamination (PNWS–AWWA 1996). Table B.10 summarizes some examples of waterborne disease outbreaks due to distribution system infrastructure failure. Table B.11 summarizes characteristics of common waterborne pathogens.

Chemical Contaminants

As with microbial contaminants, chemicals can enter the water distribution system from the following sources:

- Source water (i.e., geological quality, influence of upstream users, pollutants, etc.)
- In-process chemical
- Backsiphonage and cross-connections within the distribution system
- Infiltration from pipe bedding material during periods of low distribution system pressure

Potable water regulations require water utilities to monitor and control chemical pollutants based on regulations and guidelines governing safe and/or acceptable levels of specific chemicals in drinking water (Cohn, Cox, and Berger 1999).

Acceptable and/or safe levels of chemicals are determined through toxicological studies of route of exposure, dose, and response (Amdur, Doull, and Klaassen 1991). Dose–response relationships can vary according to the age of a person. Young children and neonates typically show a significant response to very low doses, whereas the elderly, with significantly lower metabolic rates, would exhibit a lesser response to the same dose (Klassen and Rozman 1991). Health effects due to chemical exposures are typically described as

- Toxic: causing harmful response, seriously disrupting function, or causing death
- Carcinogenic: causing or inducing uncontrolled cell growth into malignant tumors
- Genotoxic: causing alteration to genetic cell materials (e.g., chromosome alteration, unregulated DNA synthesis)
- Mutagenic: causing heritable change in the genetic code of living cells
- Teratogenic: causing birth defects (Cohn, Cox, and Berger 1999; WHO 1996)

As shown in Table B.1, there are numerous cases of chemical contamination of the distribution system from backsiphonage, especially from herbicides and pesticides. In the majority of cases, long-term health assessment of exposure was not done (PNWS–AWWA 1996). Hydraulic

Table B.11
Common waterborne disease pathogens

Pathogen	Pathogenicity		Route of transmission			Typical incubation period	Primary disease caused*	Potential sources
	Opportunistic	Direct	Food	Water	Person to person			
Bacteria								
Aeromonas hydrophila	+	+	+	+		NA†	Gastroenteritis, septicemia	Soil, vegetation, surface water, groundwater, recreational water, sewage, distribution system (DS) biofilms
Bacillus cereus	+	+	+	+			Gastroenteritis	Soil, surface water, sewage, humans
Campylobacter coli	+	+	+	+	+	3 to 5 days	Gastroenteritis	Soil, vegetation, surface water, recreational water, sewage, humans, animals
Campylobacter jejuni	+	+	+	+	+	3 to 5 days	Gastroenteritis	Soil, vegetation, surface water, recreational water, sewage, humans, animals
Citrobacter freundii	+		+	+	+		Gastroenteritis	Soil, vegetation, surface water, recreational water, sewage, humans, animals, DS biofilms
Clostridium perfringens	+		+	+			Gastroenteritis	Soil, vegetation, surface water, recreational water, sewage, humans, animals
Enterobacter aerogenes	+			+	+	10 to 72 hours	Gastroenteritis	Soil, vegetation, surface water, recreational water, DS biofilms, sewage, humans
Enterococcus fecium	+			+	+	10 to 72 hours	Gastroenteritis	Soil, vegetation, surface water, recreational water, DS biofilms, sewage, humans, animals
Escherichia coli (hemorrhagic)		+	+	+	+	12 to 60 hours	Gastroenteritis	Soil, vegetation, surface water, sewage, humans, animals
E. coli (toxigenic)		+	+	+	+	9 to 12 hours	Gastroenteritis	Soil, vegetation, surface water, sewage, humans, animals
E. coli	+		+	+	+	10 to 72 hours	Gastroenteritis	Soil, vegetation, surface water, recreational water, sewage, DS biofilms, humans, animals
Klebsiella pneumophila	+		+	+	+		Gastroenteritis	Soil, vegetation, surface water, recreational water, sewage, DS biofilms, humans

(continued)

Table B.11 (Continued)

Pathogen	Pathogenicity		Route of transmission			Typical incubation period	Primary disease caused*	Potential sources
	Opportunistic	Direct	Food	Water	Person to person			
Legionella pneumophila	+	+		+	+	2 to 10 days; 24 to 48 hours for Pontiac fever	Legionnaires' disease, Pontiac fever	Soil, surface water, groundwater, air conditioners, humidifiers, hot-water tanks, recreational water, humans
Mycobacterium avium-intracellulare	+	+		+	+		Pneumonia, septicemia	Soil, surface water, recreational water, sewage, food, birds
Pseudomonas aeruginosa	+	+		+	+		Gastroenteritis, pneumonia, folliculitis	Soil, vegetation, surface water, recreational water, food, humans
Pseudomonas cepecia	+			+			Pneumonia	Soil, surface water, recreational water, DS biofilms, humans
Salmonella species		+	+	+	+	6 to 72 hours	Salmonellosis, gastroenteritis	Soil, vegetation, surface water, recreational water, food, humans, animals
Shigella species		+	+	+	+	12 to 96 hours	Shigellosis, gastroenteritis	Soil, vegetation, surface water, recreational water, food, humans, animals
Staphylococcus aureus	+	+	+	+	+	Variable; typically 4 to 10 days	Gastroenteritis, septicemia, wound infection	Soil, vegetation, surface water, recreational water, food, humans, animals
Vibrio cholerae		+		+	+	A few hours to 5 days	Cholera	Soil, surface water, groundwater, recreational water, food, humans, shellfish
Yersinia enterocolitica	+		+	+		3 to 10 days	Yersiniosis, enterocolitis	Soil, vegetation, surface water, recreational water, food, humans, animals

(continued)

Table B.11 (Continued)

Pathogen	Pathogenicity		Route of transmission			Typical incubation period	Primary disease caused*	Potential sources
	Opportunistic	Direct	Food	Water	Person to person			
Amoeba								
Acanthamoeba species	+			+		3 to ≥7 days	Conjunctivitis, skin infection, meningo-encephalitis	Soil, vegetation, surface water, recreational water, food, humans, animals
Naegleria fowlerii	+			+		3 to 7 days	Meningo-encephalitis	Hot springs, hot tubs, sewage
Viruses								
Adenovirus	+			+	+		Conjunctivitis, gastroenteritis, pneumonia	Surface water, groundwater, recreational water, sewage, humans
Calicivirus	+			+	+		Gastroenteritis	Surface water, groundwater, recreational water, sewage, humans, animals, birds
Enterovirus	+		+	+	+	24 to 7 hours	Gastroenteritis	Surface water, groundwater, recreational water, sewage, humans, animals, birds
Hepatitis A	+	+	+	+	+	15 to 5 days	Gastroenteritis	Surface water, groundwater, recreational water, sewage, humans, shellfish
Norwalk virus	+	+		+	+	24 to 48 hours	Gastroenteritis	Surface water, groundwater, recreational water, sewage, humans, animals, birds
Rotavirus	+			+	+	24 to 72 hours	Gastroenteritis	Surface water, groundwater, recreational water, sewage, humans, animals, birds
Polio virus		+		+	+	3 to 35 days (7 to 14 days, reported range)	Poliomyelitis	Surface water, groundwater, recreational water, sewage, humans, animals, birds

(continued)

Table B.11 (Continued)

Pathogen	Pathogenicity		Route of transmission			Typical incubation period	Primary disease caused*	Potential sources
	Opportunistic	Direct	Food	Water	Person to person			
Protozoa								
Balintinium coli		+	+	+	+	Unknown	Balantidial dysentery	Surface water, groundwater, recreational water, sewage, humans, animals,
Cryptosporidium parvum	+	+	+	+	+	1 to 12 days	Cryptosporidiosis	Surface water, groundwater, recreational water, sewage, humans, animals, birds
Cyclospora cayetanensis		+	+	+	+		Gastroenteritis	Surface water, recreational water, sewage, humans
Entamoeba histolytica	+			+	+	Variable; typically 2 to ≥4 weeks	Dysentery	Surface water, recreational water, sewage, humans
Giardia lamblia	+	+	+	+	+	5 to ≥25 days	Giardiasis	Surface water, groundwater, recreational water, sewage, humans, animals
Toxoplasma gondii	+	+		+	+	5 to 20 days	Toxoplasmosis	Surface water, groundwater, recreational water, sewage, humans, animals

Source: Adapted from Benenson 1995, Koneman et al. 1988, Lennette et al. 1985, Mims et al. 1993.
*Disease caused is a function of route of entry (i.e., ingestion, inhalation, dermal, etc.), infectious dose received, health of the individual (i.e., immunocompromised, immunodeficient, immunosuppressed), and risk group potentially exposed (i.e., children and infants, adults, elderly, the ill, the institutionalized, etc.).
†NA = no information available.

modeling can be used successfully to follow contaminant propagation throughout the distribution system, and can be especially useful during waterborne disease outbreaks to pinpoint sources of contaminants (Clark and Grayman 1998).

During the period from 1971 to 1998, chemical intoxication was responsible for approximately 39 percent of waterborne disease outbreaks in the United States (Craun and Calderon 2001). Copper intoxication was the most frequent, followed by other chemicals (fluoride, lead, chromium, sodium hydroxide, chlorine, liquid soap, ethyl acrylate, morpholine, hydroquinone—1 outbreak each), chlordane, nitrite, ethylene glycol, an unidentified herbicide, and oil (Craun and Calderon 2001). Most of these outbreaks were the result of cross-connections and backsiphonage to the distribution system. The outbreaks were the result of corrosion of the plumbing system, lack of backflow devices, improperly maintained backflow devices, or improperly installed backflow devices (Craun and Calderon 2001). One outbreak in 1997 was the result of improper grounding of a building's electrical wiring to a restaurant's plumbing system (Barwick et al. 2000), wherein two persons became ill after ingesting a fruit drink made with restaurant tap water (Barwick et al. 2000).

CONCLUSIONS

Water distribution systems play a key role in the transmission of biological and chemical contaminants to community residents and businesses. Failure to protect the distribution system from contaminant intrusion, whether by improper installation, improper maintenance, high frequency of main breaks, or other reasons, ensures that the community will have numerous opportunities for contaminant exposure and subsequent health effects.

Protection of public health from waterborne disease requires the collaborative efforts of water utilities, public health agencies, and regulatory bodies. An active utility surveillance system, working in concert with an active health surveillance system, will serve to provide the best guarantee of drinking water safety for a community.

Distribution system infrastructure is vulnerable to innumerable breaches of functional integrity. Utilities and health agencies located in known risk areas (i.e., seismic zones, etc.) need to have up-to-date emergency response plans to minimize the effects of known natural disasters. The same is true for utilities and public health agencies in low-risk or no-risk areas. Human-made disasters, intentional or unintentional, can occur at any time and anywhere. Planning for emergencies ensures prompt response and remedial actions. This serves to protect drinking water quality and community health.

APPENDIX C
POTENTIAL RESEARCH FUNDING RESOURCES

Funding for fundamental research and/or specific infrastructure failure research and associated health effects is becoming increasingly scarce and competitive. Irrespective of size, municipalities continue to be faced with shrinking budgets to accommodate ever-expanding needs, including water infrastructure. Research and infrastructure funding typically remains localized at regional and individual levels. This is a fundamental function of taxation and a fee-for-service funding base that is available to water utilities and/or public health departments.

AwwaRF MODEL

The Awwa Research Foundation (AwwaRF) is one model for fundamental research into issues of concern for water utilities and public health agencies. Funding is derived from subscriber utilities, with a research agenda that is determined on an annual basis, and which is responsive to the needs of industry and government. AwwaRF is one of the few nonpartisan, fundamental, international research agencies available for drinking water research. Research partners include water utilities throughout North America, Europe, and Australia, as well as complementary academic institutions, public health institutions, and industrial partners.

LOCAL AND/OR REGIONAL MODELS

Many examples of effective partnerships, within and external to the water industry, exist and continue to serve as a reminder of the need to build strong community relationships to achieve common objectives. Some of these partnerships function at an international level with respect to funding, but actually operate at a local level with respect to implementation. One example of this is the American Water Works Association's Water For People program. Other examples include collaborative funding programs through the World Bank, the World Health Organization, and the United Nations, where funding is targeted to specific communities but the solution involves multiple international partners.

Multiple local partnerships exist throughout the developed world. This includes partnerships between local municipal agencies together with provincial/state agencies, and with federal agencies and/or academic institutions. Funding for these projects can be derived from a multiplicity of resources, including all levels of government involved. As E. Geldreich shared during his 2002 workshop discussion, if the issue was important enough to the community, the community would find a way to fund a solution. The grassroots influence cannot be discounted when trying to find new sources of funding.

Collaboration with other stakeholder agencies will continue to be critical to sourcing new, innovative resources. As discussed previously, this requires open, active, honest, and mutual efforts at problem definition and resolution. It requires clear definition of mutual interests, strengths, and also weaknesses. It requires a common goal and purpose—to minimize community waterborne disease by minimizing infrastructure failures.

REFERENCES

Alber K., A. Bruchet, and E. Shpirt. 1989. *Leachate From Organic Coating Materials Used in Potable Water Distribution Systems*. Denver, Colo.: AWWA.

Allen, M.J., J.L. Clancy, and E.W. Rice. 2000. The Plain, Hard Truth About Pathogen Monitoring. *Journal AWWA*, 92(9):64–76.

Amann, R.I., H.C. Glemming, U.H.G. Obst, L. Bartolucci, A. Pariani, M.E. Guerzoni, F. Westall, F. Gardini, H. Abouzaid, M. El Mghari Tabib, A. Outair, L. Laraki, and S. Benabdallah. 1996. Interaction by Microbiological Processes Between Water, Biofilm, and Pipe Materials in Water Distribution Systems. *Water Supply*, 14(3–4):453–471.

Amdur, M.O., J. Doull, and C.D. Klaassen, eds. 1991. *Casarett and Doull's Toxicology*. 4th ed. New York: Pergamon Press.

Anne Anderson et al. v. Cryovac et al. 1983. Case No. 82-1672-S. U.S. District Court for the First Circuit, Boston, Mass.

Anon. 2002. Seismic Intensity Scales [Online]. Available: <http://utenti.lycos.it/unita_di_misura/sism_e/htm>. [cited April 1, 2002]

Antoun, E.N., R.J. Theiss, J.E. Dyksen, and D. Hiltebrand. 1995. Role of Distribution Systems in Achieving Compliance With Safe Drinking Water Act. In *Proc. of the AWWA Annual Conference*. Denver, Colo.: AWWA.

AOAC International. 2002. *Official Methods of Analysis of AOAC International* [Online]. Available: <http://www.aoac.org/pubs>. [cited October 7, 2003]

APHA, AWWA, and WEF (American Public Health Association, American Water Works Association, and Water Environment Federation). 2005. *Standard Methods for the Examination of Water and Wastewater*. 21st ed. Washington, D.C.: APHA.

APHA, AWWA, and WPCF (American Public Health Association, American Water Works Association, and Water Pollution Control Federation). 1976. *Standard Methods for the Examination of Water and Wastewater*. 14th ed. Washington, D.C.: APHA.

Arastu, A.H., L.D. Noble, W.L. LaFramboise, and J.E. Rhoads. 1999. Diagnostic Evaluation of a Severe Water Hammer Event in the Fire Protection System of a Nuclear Power Plant. In *Proc. of the Third ASME/JSME Fluids Engineering Conference*. New York: American Society of Mechanical Engineers.

Arquette, M., M. Colk, K. Cook, B. LaFrance, M. Peters, J. Ransom, E. Sargent, V. Smoke, and A. Stairs. 2002. Holistic Risk-Based Environmental Decision Making: A Native Perspective. *Environmental Health Perspectives*, 110(supplement 2):259–264.

ASQ (American Society for Quality). 2003. *W. Edwards Deming: A Mission Pursued on Two Continents* [Online]. Available: <http://www.asq.org/join/about/history/deming.html>. [cited May 14, 2003]

ASTDHPPHE (Association of State and Territorial Directories of Health Promotion and Public Health Education). 2000. *Emergency Response Communications Plans for Infectious Disease Outbreaks and Bioterrorist Events*. Washington, D.C.: ASTDHPPHE.

auf der Heide, E. 1989. *Disaster Response: Principles of Preparation and Coordination*. St. Louis, Mo.: C.V. Mosby.

Aurelius, E., ed. 1994. *The January 17, 1994 Northridge, CA Earthquake*. An EQE Summary Report. San Francisco: EQE International.

Australian Food Secretariat. 2000. *Principles and Operating Protocols for Stakeholder Consultation* [Online]. Available: <http://www.foodsecretariat.health.gov.au>. [cited March 28, 2003]

AWWA (American Water Works Association). 1986a. *Introduction to Water Distribution*. Denver, Colo.: AWWA.

AWWA (American Water Works Association). 1986b. *Maintaining Distribution System Water Quality*. Denver, Colo.: AWWA.

AWWA (American Water Works Association). 1994. *Emergency Planning for Water Utilities*. Denver, Colo.: AWWA.

Babbit, H.E., and J.J. Doland. 1949. *Water Supply and Engineering*. New York: McGraw-Hill.

Bartlett, A.V., M. Moore, G.W. Gary, K.M. Starko, J.J. Erben, and B.A. Meredith. 1985. Diarrheal Illness Among Infants and Toddlers in Day Care Centers. I. Epidemiology and Pathogens. *Journal of Pediatrics*, 197(4):495–502.

Bartram, J. 1999. Emerging Issues on Waterborne Pathogens. In *Proc. of the AWWA International Symposium on Waterborne Pathogens*. Denver, Colo.: AWWA.

Barwick, R.S., D.A. Levy, G.F. Craun, M.J. Beach, and R.L. Calderon. 2000. Surveillance for Waterborne-Disease Outbreaks—United States, 1997–1998. *Morbidity and Mortality Weekly Report*, 49:4,1.

Basch, P.F. 1990. *Textbook of International Health*. New York: Oxford University Press.

BEHA (Bureau of Environmental Health Assessment). 1997. *Woburn Childhood Leukemia Follow-up Study* [Online]. Available: <http://www.state.ma.us/dph/beha/cau/reports/woburn/leukfact.htm>. [cited April 25, 2003]

Bendell, T. 1998. *The Quality Gurus: Help or Hype for British Industry?* [Online] Available: <http://dtiinfo.1.dti.gov.uk/mbp/bpgt/m9ja00001>. [cited May 13, 2003]

Benenson, A.S., ed. 1995. *Control of Communicable Diseases Manual*. 16th ed. Washington D.C.: American Public Health Association.

Bernstein, P.L. 1996. The New Religion of Risk Management. *Harvard Business Review*, (March–April): 47.

Binder, S., and L.M. Sanderson. 1987. The Role of the Epidemiologist in Natural Disasters. *Annals of Emergency Medicine*, 16(9):1081–1084.

Birch, S., G. Stoddart, and F. Beland. 1998. Modelling the Community as a Determinant of Health. *Canadian Journal of Public Health*, 89(6):402–405.

Bissell, R.A. 1983. Delayed-Impact Infectious Disease After A Natural Disaster. *Journal of Emergency Medicine*, 1:59–63.

Blake, P.A., M.L. Rosenbern, J.B. Costa, P.S. Ferreira, C.L. Guimaraes, and E.J. Gangarosa. 1977. Cholera in Portugal, 1974. II. Transmission by Bottled Water. *American Journal of Epidemiology*, 105:344–348.

Blatt, S.P., M.D. Parkinson, E. Pace, P. Hoffman, D. Dolan, P. Lauderdale, R.A. Zajac, and G.P. Melcher. 1993. Nosocomial Legionellosis: A One Year Case Control and Environmental Investigation. In *Legionella: Current Status and Emerging Perspectives*. Edited by J.M. Barbaree, R.F. Breiman, and A.P. Dufour. Washington, D.C.: American Society for Microbiology.

Block, J.C., M. Dutang, J. Maillard, and D. Reasoner. 1994. Growth of Attached Bacteria in Water Distribution Systems. *Water Supply*, 12:SS 1-8–1-11.

Borenstein, S.W. 1994. *Microbially Influenced Corrosion Handbook*. New York: Industrial Press.

Boswell, T.C., D.T. Darnell, J.L. Ledbetter, J. Benson, T.W. McKinley, J.D. Smith, R.K. Sikde, and H.B. Greenberg. 1982. Community Outbreak of Norwalk Gastroenteritis—Georgia. *Morbidity and Mortality Weekly Report*, 31(30):405–407.

Boutilier, M.A., R.F. Badgley, B.D. Poland, and S. Tobin. 2001. Playing on Shifting Sand: Reflections on the Impact of Political Shifts on Community Action and Public Health. *Canadian Journal of Public Health*, 92(2):87–89.

Bradford, S.A. 1993. *Corrosion Control*. New York: Van Nostrand Reinhold.

Breiman, R.F. 1993. Modes of Transmission in Epidemic and Nonepidemic *Legionella* Infection: Directions for Further Study. In *Legionella: Current Status and Emerging Perspectives*. Edited by J.M. Barbaree, R.F. Breiman, and A.P. Dufour. Washington, D.C.: American Society for Microbiology.

Brown, P. 1992. Popular Epidemiology and Toxic Waste Contamination: Lay and Professional Ways of Knowing. *Journal of Health and Social Behaviour*, 33(9):267–281.

Bull, R.J., and R.D. Bull. 2001. Risk Assessment. In *Health Effects and Occurrence of Disinfection By-Products*. Edited by E.J. Bull, S.W. Krasner, P.A. Daniel, and R.D. Bull. Denver, Colo.: AwwaRF and AWWA.

Burlingame, G.A., J. Rahman, E. Navera, and J.E. Durrant. 1998. Managing Water Supply Infrastructure. *Journal AWWA*, 90(7):53–61.

Burns, R., F. Darr, J. Wilhelm, and V.W. Davis. 2001. National Security and State Public Utility Commissions. *Security Issues. The National Regulatory Research Institute* [Online]. Available: <http://www.nrri.org/download.php?id=02-17>. [cited September 20, 2001]

The Business Journal. 2002. Recent Water Main Breaks Hurt Business, Raise Infrastructure Issues [Online]. Available: <http://bizjournals.com/kansascity/stories/2000/10/16/focus4)>. [cited April 1, 2002]

Calderon, R.L., C.C. Johnson Jr., G.F. Craun, A.P. Dufour, R.J. Karlin, T. Sinks, and J.L. Valentine. 1993. Health Risks From Contaminated Water: Do Class and Race Matter? *Toxicology and Industrial Health*, 9(5):879–900.

Cameron, C., and C. James. 2002. *The 1923 Great Kanto Earthquake and Fire*. National Information Service for Earthquake Engineering. University of California–Berkeley [Online]. Available: <http://nisee.berkeley.edu.kanto/yokohama.html>. [cited March 18, 2002]

Camper, A.K., M. Warnecke, W.L. Jones, and G.A. McFeters. 1998. *Pathogens in Model Distribution System Biofilms*. Denver, Colo.: AwwaRF and AWWA.

Carmichael, A.G. 1993a. Diseases of the Renaissance and Early Modern Europe. In *The Cambridge World History of Human Disease*. Edited by K.F. Kiple. New York: Cambridge University Press.

Carmichael, A.G. 1993b. History of Public Health and Sanitation in the West Before 1700. In *The Cambridge World History of Human Disease*. Edited by K.F. Kiple. New York: Cambridge University Press.

CDC (Centers for Disease Control and Prevention). 2002. *Local Emergency Preparedness and Response Inventory: A Tool for Rapid Assessment of Local Capacity to Respond to Bioterrorism, Outbreaks of Infectious Disease, and Other Public Health Threats and Emergencies*. Public Health Program Office. Atlanta, Ga.: CDC.

Cesario, L. 1995. *Modeling, Analysis, and Design of Water Distribution Systems*. Denver, Colo.: AWWA.

Chadderton, R.A., G.L. Christensen, and P. Henry-Unrath. 1992. *Implementation and Optimization of Distribution Flushing Programs*. Denver, Colo.: AwwaRF and AWWA.

Chalecki, E.L. 2001. *A New Vigilance: Identifying and Reducing the Risks of Environmental Terrorism*. A Report of the Pacific Institute for Studies in Development, Environment and Security, Oakland, Calif. [Online]. Available: <http://www.pacinst.org/environmental_ terrorism_final_pdf>. [cited May 20, 2002]

Chang, S.L., G. Berg, N.A. Clarke, and P.W. Kabler. 1960. Survival and Protection Against Chlorination of Human Pathogens in Free-Living Nematodes Isolated From Water Supplies. *American Journal of Tropical Medicine and Hygiene*, 9:136–142.

Characklis, W.G., and K.E. Cooksey. 1983. Biofilms and Microbial Fouling. *Advances in Applied Microbiology*, 29:93–138.

Chartier, J., and S. Gabler. 2001. *Risk Communication and Government: Theory and Application for the Canadian Food Inspection Agency* [Online]. Available: <http://www.inspection.gc.ca/english>. [cited April 1, 2003]

Chow, C.W., L.R. Lee-Oack, N. Senathiragah, M. Rawji, M. Chan, and C.K. Chan. 1995. Community Acquired, Nursing Home Acquired and Hospital Acquired Pneumonia: A Five-Year Review of the Clinical, Bacteriological and Radiological Characteristics. *Canadian Journal of Infectious Diseases*, 6(6):317–325.

Chronic Diseases Division (Center for Environmental Health). 1981. *Memo Regarding Cancer in Woburn Massachusetts* [Online]. Available: <http://www.state.ma.us/dph/beha/cau/reports/woburn>. [cited May 1, 2003]

Clark, R.M., and R.A. Deininger. 2000. Protecting the Nation's Critical Infrastructure: The Vulnerability of U.S. Water Supply Systems. *Journal of Contingencies and Crisis Management*, 8(2):73–80.

Clark, R.M., and R.A. Deininger. 2001. Minimizing the Vulnerability of Water Supplies to Natural and Terrorist Threats. In *Proc. of the AWWA IMTech Conference*. Denver, Colo.: AWWA.

Clark, R.M., E.E. Geldreich, K.R. Fox, E.W. Rice, C.H. Johnson, J.A. Goodrich, J.A. Barnick, and F. Abdesaken. 1996a. Tracking a *Salmonella* serovar *typhimurium* Outbreak in Gideon, Missouri: Role of Contaminant Propagation Modelling. *Journal of Water Supply: Research and Technology–Aqua*, 45(4):171–183.

Clark, R.M., E.E. Geldreich, K.R. Fox, E.W. Rice, C.W. Johnson, J.A. Goodrich, J.A. Barnick, F. Abdesaken, J.E. Hill, and F.J. Angulo. 1996b. A Waterborne *Salmonella typhimurium* Outbreak in Gideon Missouri: Results From a Field Investigation. *International Journal of Environmental Health Research*, 6:187–193.

Clark, R.M., and W.M. Grayman. 1998. *Modeling Water Quality in Drinking Water Distribution Systems*. Denver, Colo.: AWWA.

Cloud, W.K., D.M. Hill, M.E. Huffman, C.W. Jennings, T.V. McEvilly, R.D. Nason, K.V. Tocher D. Steinbrugge, J.D. Unger, and T.L. Youd. 2002. *The 1969 Santa Rosa Earthquake* [Online]. Available: <http://www.johnmartin.com/earthquakes/eqpapers/0000001>. [cited March 1, 2002]

Cohn, P.D., M. Cox, and P.S. Berger. 1999. Health and Aesthetic Aspects of Water Quality. In *Water Quality and Treatment*. Edited by R.D. Letterman. New York: McGraw-Hill.

Cohrssen, J.J., and V.T. Covello. 1989. *Risk Analysis: A Guide to Principles and Methods for Analyzing Health and Environmental Risks*. Washington, D.C.: U.S. Council on Environmental Quality, Executive Office of the President.

Conroy, P.J., D.M. Hughes, and I. Wilson. 1995. *Demonstration of an Innovative Water Main Rehabilitation Technique: In Situ Epoxy Lining*. Denver, Colo.: AwwaRF and AWWA.

Costerton, J.W., and J. Boivin. 1991. The Role of Biofilms in Microbial Corrosion. In *Microbially Influenced Corrosion and Biodeterioration*. Edited by N.J. Dowling, M.W. Mittelman, and J.C. Danko. Knoxville, Tenn.: University of Tennessee Press.

Costerton, J.W., K.J. Cheng, G.G. Geesey, T.I. Ladd, J.C. Nickel, M. Dasgupta, and T.J. Marrie. 1987. Bacterial Biofilms in Nature and Disease. *Annual Review of Microbiology*, 41:435–464.

Craun, G.F., P.B. Berger, and R.L. Calderon. 1997. Coliform Bacteria and Waterborne Disease Outbreaks. *Journal AWWA*, 89(3):96–194.

Craun, G.F., and R.L. Calderon. 2001. Waterborne Disease Outbreaks Caused by Distribution System Deficiencies. *Journal AWWA*, 93(9):64–75.

Cromwell, J., G. Nestel., R. Albani, L. Paralez, A.K. Deb, F.M. Grablutz. 2001. *Financial and Economic Optimization of Water Main Replacement Programs*. Denver, Colo.: AwwaRF and AWWA.

Davies, C., D.L. Fraser, P.C. Hertzler, and R.T. Jones. 1997. USEPA's Infrastructure Needs Survey. *Journal AWWA*, 89(12):30–38.

Dawe, J.L., C.K. Seah, and A.K. Abdel-Zaher. 1993. Investigation of the Regent Street Water Tower Collapse. *Journal AWWA*, 85(5):34–47.

Dean, E.B. 2003. Risk From the Perspective of Competitive Advantage [Online]. Available: <http://www.dfca.org/sys/rsk.html>. [cited January 12, 2003]

Deb, A.K., F.M. Grablutz, Y.J. Hasit, R.K. Herz, Z.S. Moumne. 1997. Water Main Rehabilitation Forecasting Model and Its Application. In *Proc. of the AWWA Annual Conference*, Vol. E. Denver, Colo.: AWWA.

Deb, A.K., Y.J. Hasit, and F.M. Grablutz. 1995. *Distribution System Performance Evaluation*. Denver, Colo.: AwwaRF and AWWA.

Deb, A.K., Y.J. Hasit, F.M. Grablutz, and R.K. Herz. 1998. *Quantifying Future Rehabilitation and Replacement Needs of Water Mains*. Denver, Colo.: AwwaRF and AWWA.

Deb, A.K., K.A. Momberger, Y.J. Hasit, and F. Grablutz. 2000. *Guidance for Management of Distribution System Operation and Maintenance*. Denver, Colo.: AwwaRF and AWWA.

Delvecchio, F. 1998. Using Need-To-Know Criteria to Design Training. *Journal AWWA*, 90(10):56–59.

Denileon, G.P. 2001. The Who, What, Why, and How of Counterterrorism Issues. *Journal AWWA*, 93(5):78–85.

Disaster Relief. 1999. Mexico Earthquake–Oaxaca State [Online]. Available: <http://www.disasterrelief.org/Disasters/990930mcquake2/>. [cited March 1, 2002]

DoD (U.S. Department of Defense). 1999. *DoD Y2K Contingency Planning* [Online]. Available: <http://www.c#i.osd.mil/org/cio/y2k/y2k_con_plan/>. [cited February 15, 2002]

Doggett, M.S. 2000. Characterization of Fungal Biofilms Within a Municipal Water Distribution System. *Applied and Environment Microbiology*, 66(3):1249–1251.

Duffy, J. 1993. History of Public Health and Sanitation in the West Since 1700. In *The Cambridge World History of Human Disease*. Edited by K.F. Kiple. New York: Cambridge University Press.

du Moulin, G.C., K.D. Stottmeier, P.A. Pelletier, A.Y. Tsang, and J. Hedley-Whyte. 1988. Concentration of *Mycobacterium avium* by Hospital Hot Water Systems. *Journal of the American Medical Association*, 260:1599–1601.

Ehrlich, H.L. 1990. *Geomicrobiology.* New York: Marcel Dekker.

Eisenberg, J.N.S., M.A. Brookhart, G. Rice, M. Brown, and J.M. Colford Jr. 2002. Disease Transmission Models for Public Health Decision Making: Analysis of Epidemic and Endemic Conditions Caused by Waterborne Pathogens. *Environmental Health Perspectives*, 110(8):783–790.

Eliassen, R., and R.H. Cummings. 1948. Analysis of Water-Borne Outbreaks, 1938–1945. *Journal AWWA*, 40:509–528.

Elkhamri, M., L.W. Grau, L. King-Irani, A.S. Mitchell, and L. Tasa-Bennett. 2005. Urban Population Control in a Counterinsurgency [Online]. Available: <http://www.stormingmedia.us/38/3894/A389434.html> [cited May 27, 2006]

Ellison, D., A. Fomer, G. Bell, and A. O'Brian. 2001. *Distribution Infrastructure Management: Answers to Common Questions.* Denver, Colo.: AwwaRF and AWWA.

Ellison, R.J., R.C. Linger, H.F. Lipson, N.R. Mead, and A. Moore. 2002. Foundations for Survivable Systems Engineering [Online]. Availble: <http://www.cert.org.archive/html/SSE-foundations.html>. [cited June 2, 2002]

Emde, K., L. Gammie, N. Fok, A. Barry, J. Mainiero, B. Reilley-Matthews, and J. Talbot. 2001. Case Study—A Testing of Two Cities. In *Guidance Manual for Waterborne Gastrointestinal Disease Outbreak Detection.* Edited by K.M.E. Emde, J.A. Talbot, L. Gammie, J.M. Mainiero, E.E. Geldreich, M.A. Barry, and N. Fok. Denver, Colo.: AwwaRF and AWWA.

Emde, K., E. Geldreich, and J. Talbot. 2001. Introduction to Safe Drinking Water. In *Guidance Manual for Waterborne Gastrointestinal Disease Outbreak Detection.* Edited by K.M.E. Emde, J.A. Talbot, L. Gammie, J.M. Mainiero, E.E. Geldreich, M.A. Barry, and N. Fok. Denver, Colo.: AwwaRF and AWWA.

Emde, K.M., D.W. Smith, and W. Albritton. 1994. Potential Health Concerns Arising From Water Distribution System Corrosion. In *Proc. of the Municipal and Rural Water Supply and Water Quality Conference.* Poznan, Poland: Polish Association of Sanitary Engineers and Technicians.

Emde, K.M.E., and D.W. Smith. 1992. *Significance of Microbially Influenced Corrosion on Water Distribution Systems.* Unpublished report prepared for Alberta Environment, Edmonton, Alta.

Emde, K.M.E., D.W. Smith, and S.J. Stanley. 1995. *Water Distribution Flushing and Cast Iron Pigging Study.* Unpublished report prepared for Water Network Maintenance, Department of Public Works, City of Edmonton, Edmonton, Alta.

EQE Engineering Inc. 1989. Loma Prieta Earthquake [Online]. Available: <http://www.eqe.com/publications/lomatrie/lomaprie.htm>. [cited March 1, 2002]

Evans, A.S. 1993. *Causation and Disease: A Chronological Journey.* New York: Plenum Medical Book Company.

Evans, R., and G. Stoddart. 1990. Producing Health, Consuming Health Care. *Social Science Medicine*, 31(12):1347–1363.

FAA (United States Federal Aviation Administration). 2003. [Online]. Available: <http://www1.faa.gov>. [cited April 1, 2003]

Fay, M. 2001. Putting the Public Back Into Public Health: One City's Experience. *Canadian Journal of Public Health*, 92(2):85–86.

Feinstein, A.R. 1985. *Clinical Epidemiology.* Philadelphia: W.B. Saunders.

FEMA (Federal Emergency Management Agency). 2000. FEMA Urges Consumers to Keep Y2K Preparedness Items [Online]. Available: <http://www.fema.gov/nwz00/consumers 0106.html>. [cited February 1, 2002]

Fink, S. 1986. *Crisis Management—Planning for the Inevitable*. New York: American Management Association.

Fischoff, B. 2003. Acceptable Risk: A Conceptual Proposal [Online]. Available: <http://www.fplc.edu/RISK/vol5/winter/Fischhof.html>. [cited January 20, 2003]

Fontana, M. 1987. *Corrosion Engineering*. 3rd ed. New York: McGraw-Hill.

Ford, T., and R. Mitchell. 1990. The Ecology of Microbial Corrosion. *Advances in Microbial Ecology*, 11:231–261.

Foster, K.R., D.W. Bernstein, and P.W. Huber, eds. 1993. *Phantom Risk: Scientific Inference and the Law*. Cambridge, Mass.: The MIT Press.

Fox, M.A., J.D. Groopman, and T.A. Burke. 2002. Evaluating Cumulative Risk Assessment for Environmental Justice: A Community Case Study. *Environmental Health Perspectives*, 110 (suppl. 2):203–209.

Gale, P. 1996. Developments in Microbiological Risk Assessment Models for Drinking Water—A Short Review. *Journal of Applied Bacteriology*, 81:403–410.

Gale, P., H. van Dijk, and G. Stanfield. 1997. Drinking Water Treatment Increases Microorganism Clustering: The Implications for Microbiological Risk Assessment. *Journal of Water Supply: Research and Technology–Aqua*, 46(3):117–126.

Garrett, L. 1994. Nosocomial Amplifiers of Microbial Emergence. *Annals of the New York Academy of Sciences*, 740:389–395.

GASB (Governmental Accounting Standards Board). 2003. *GASB34 Regulations* [Online]. Available: <http://www.gasb34.org>. [cited February 14, 2003]

Gavan, D.T., and J.W. Nutt. 1970. An Epidemic of Waterborne Infectious Hepatitis in France. *Archives of Environment Health*, 20(4):523–532.

Geldreich, E.E. 1992. Vision of the Future in Drinking Water Microbiology. *Journal of the New England Water Works Association*, CVI:1–8.

Geldreich, E.E. 1994. Microbial Water Quality Concerns in Distribution Systems: Problems and Solutions. *Revista Italiana D'Igiene*, 23(1–2):3–24.

Geldreich, E.E. 1996. *Microbial Quality of Water in Distribution Systems*. Boca Raton, Fla: Lewis Publishers.

Geldreich, E.E., and M.W. LeChevallier. 1999. Microbiological Quality Control in Distribution Systems. In *Water Quality and Treatment*. Edited by R.D. Letterman. New York: McGraw-Hill.

Gerba, C.P., J.B. Rose, and C.N. Haas. 1996. Sensitive Populations: Who Is at the Greatest Risk? *International Journal of Food Microbiology*, 30:113–123.

Glass, R.I., and E.K. Noji. 1992. Epidemiologic Surveillance Following Disasters. In *Public Health Surveillance*. Edited by W. Halperin and E.L. Baker. New York: Van Nostrand Reinhold.

Gleick, P. 2000. *Water Conflict Chronology*. Oakland, Calif.: Pacific Institute for Studies in Development, Environment, and Security.

Glicker, J.L. 1992. Convincing the Public that Drinking Water Is Safe. *Journal AWWA*, 84(1):46–51.

Gochfeld, M. 1992. Environmental Risk Assessment. In *Public Health and Preventive Medicine*. 13th ed. Edited by J.M. Last and R.B. Wallace. Toronto, Ont.: Appleton & Lange.

Gorman, A.E., and A. Wolman. 1939. Water-Borne Outbreaks in the United States and Canada, and Their Significance. *Journal AWWA*, 31:225–387.

Gouveia-Vigeant, T., and J. Tickner. 2003. *Toxic Chemicals and Childhood Cancer: A Review of the Evidence* [Online]. Available: <http://www.uml.edu/umlmainpage/umlnews/pfd/ChildhoodCancer>. [cited May 5, 2003]

Grabow, W.O.K. 1996. Waterborne Diseases: Update on Water Quality Assessment and Control. *Water SA*, 222:193–202.

Graman, P.S., G.A. Quinlan, and J.A. Rank. 1997. Nosocomial Legionellosis Traced to a Coontaminated Ice Machine. *Infection Control and Hospital Epidemiology*, 18(9):637–640.

Grayman, W.M., L.A. Rossman, C. Arnold, R.A. Deininger, C. Smith, J.F. Smith, and R. Schnipke. 2000. *Water Quality Modeling of Distribution System Storage Facilities*. Denver, Colo.: AwwaRF and AWWA.

Grigg, N.S. 2000. *Risk Management in Water Utilities* [Online]. Available: <http://www.engr.colostate.edu/ce/outreach/conferences/disasters/Risk_working_paper_4.doc>. [cited March 1, 2002]

Grigg, N.S. 2002. *Surviving Disasters in Water Utilities: Learning from Experience*. Denver, Colo.: AwwaRF and AWWA.

GVRD (Greater Vancouver Regional District). 2002. *Template for Water Utilities Emergency Plan* [Online]. Available: <http://www.gvrd.bc.ca/jelc>. [cited April 10, 2002]

Hamilton, M. 1996. Liquid Assets, Pure and Simple. *Washington Post*. Sept. 14, 1996.

Harr, J. 1995. *A Civil Action*. New York: Random House.

Health Canada. 2002. *From Source to Tap: The Multi-Barrier Approach to Safe Drinking Water*. [Online]. Available: <http://www.hc-sc.gc.ca/ewh-semt/pubs/water-eau/tap-source-robinet/index_e.html>. [cited October 15, 2003]

Hebert, R.E. 2001. *A Brief Discussion of Water Security Issues Following the September 11, 2001 Terrorist Attacks*. The National Council for Public-Private Partnerships [Online]. Available: <http://www.ncppp.org/inthenews/waterdiscussion.html>. [cited April 4, 2002]

Hurd, R.E. 1993. *Consumer Attitude Survey on Water Quality Issues*. Denver, Colo.: AwwaRF and AWWA.

IAMFES (International Association of Milk, Food and Environmental Sanitarians) 1996. *Procedures to Investigate Waterborne Illness*. 2nd ed. Des Moines, Iowa: IAMFES.

Information Systems and Technology, Emergency Management. 2003. Contingency and Recovery Planning: Checklist for Information Systems [Online]. Available: <http://ist-socrates.berkeley.edu:2001/em/checklist.html>. [cited December 2, 2003]

ISO (International Organization for Standardization). 2003a. About ISO—Introduction [Online]. Available: <http://www.iso.ch/iso/aboutiso>. [cited April 2, 2003]

ISO (International Organization for Standardization). 2003b. ISO 9000/ISO 14000 [Online]. Available: <http://www.iso.org/iso/en/iso9000-14000/index.html>. [cited April 2, 2003]

ISO (International Organization for Standardization). 2004. Drinking Water Supply Services: Guidelines for the Management and Assessment of the Service. Rev. 5, draft 2. [Online]. Available: <http://www.hdzv.hr/hdzv_tekst/pdf_hdzv/wg3.pdf>. [cited April 2, 2003]

Jekel, J.F., J.G. Elmore, and D.L. Katz. 1996. *Epidemiology, Biostatistics, and Preventive Medicine*. Philadelphia: W.B. Saunders.

John Martin & Associates Inc. 2002. *1985 Mexico City Earthquake* [Online]. Available: <http://www.johnmartin.cm/earthquakes/eqshow/647003_00.html>. [cited March 1, 2002]

Joly, G.S., S.M. Burgridge, R.D. Kuehn, D.J.M. Haldane, and T.J. Marrie 1993. Phenotypic and Genotypic Variability of *Legionella pneumophila* Populating Potable Waters and Patients of a Tertiary-Care Hospital Over a Five-Year Period. In *Legionella: Current Status and Emerging Perspectives.* Edited by J.M. Barbaree, R.F. Breiman, and A.P. Dufour. Washington, D.C.: American Society for Microbiology.

Joly, J.R., and M. Alary. 1993. Occurrence of Nosocomial Legionnaires' Disease in Hospitals With Contaminated Potable Water Supply. In *Legionella: Current Status and Emerging Perspectives.* Edited by J.M. Barbaree, R.F. Breiman, and A.P. Dufour. Washington, D.C.: American Society for Microbiology.

Jones, D.A. 1992. *Principles and Prevention of Corrosion.* New York: Macmillan.

Jones, I.G., and M. Roworth. 1996. An Outbreak of *Escherichia coli* O157 and Campylobacteriosis Associated With Contamination of a Drinking Water Supply. *Public Health,* 110:277–282.

Jones, J.G. 1997. The Microbiological Quality of Water: The Nature of the Problem. *Journal of Water Supply: Research and Technology–Aqua,* 46(4):346–352.

Kaplan, J.E., R.A. Goodman, L.B. Schonberger, E.C. Lippy, and G.W. Gary. 1982. Gastroenteritis Due to Norwalk Virus: An Outbreak Associated With a Municipal Water System. *Journal of Infectious Diseases,* 146(2):190–197.

Kindred, T.P. 1996. Risk Analysis and Its Application in FSIS. *Journal of Food Protection,* (Suppl.):24–30.

Kirmeyer, G.J., M. Friedman, J. Clement, A. Sandvig, P.F. Noran, K.D. Martel, D. Smith, M. LeChevallier, C. Volk, E. Antoun, D. Hiltebrand, J. Dyksen, and R. Cushing. 2000. *Guidance Manual for Maintaining Distribution System Water Quality.* Denver, Colo.: AwwaRF and AWWA.

Kirmeyer, G.J., M. Friedman, K. Martel, D. Howie, M. LeChevallier, M. Abbaszadegan, M. Karim, J. Funk, and J. Harbour. 2001. *Pathogen Intrusion Into the Distribution System.* Denver, Colo.: AwwaRF and AWWA.

Kirmeyer, G.J., L. Kirby, B.M. Murphy, P.F. Noran, K.D. Martel, T.W. Lund, J.L. Anderson, and R. Medhurst. 1999. *Maintaining Water Quality in Finished Water Storage Facilities.* Denver, Colo.: AwwaRF and AWWA.

Kirmeyer, G.J., W. Richards, and C.D. Smith. 1994. *An Assessment of Water Distribution Systems and Associated Research Needs.* Denver, Colo.: AwwaRF and AWWA.

Kirmeyer, G.J., I. Wagner, and P. Leroy. 1996. Organizing Corrosion Control Studies and Implementing Corrosion Control Strategies. In *Internal Corrosion of Water Distribution Systems.* 2nd ed. Denver, Colo.: AwwaRF and DVGW-Technologiezentrum Wasser.

Klassen, C.D., and K. Rozman. 1991. Absorption, Distribution, and Excretion of Toxicants. In *Casarett and Doull's Toxicology.* 4th ed. Edited by M.O. Amdur, J. Doull, and C.D. Klaassen. New York: Pergamon Press.

Koneman, E.W., S.D. Allen, V.R. Dowell Jr., W.N. Janda, H.M. Sommers, and W.C. Winn Jr. 1988. *Color Atlas and Textbook of Diagnostic Microbiology.* 3rd ed. Philadelphia: J.B. Lippincott.

Koplan, J.P. 2002. *Building Infrastructure to Protect the Public's Health.* Public Health Training Network Broadcast. ASTHO, Washington D.C. [Online]. Available: <http://www.cdc.gov>. [cited March 31, 2002]

Kotelchuck, M., and G.S. Parker. 1979. *Woburn Health Data Analysis 1969–1978* [Online]. Available: <http://www.state.ma.us/dph/beha/cau/reports/woburn/woo.htm>. [cited April 25, 2003]

Kramer, M.H., B.L. Herwaldt, G.F. Craun, R.L. Calderon, and D.D. Juranet. 1996. Surveillance for Waterborne-Disease Outbreaks—United States, 1993–1994. *Morbidity and Mortality Weekly Report*, 45(SS-1):1–31.

Kramer, M.H., G. Quade, P. Hartemann, and M. Exner. 2001. Waterborne Diseases in Europe—1986–1996. *Journal AWWA*, 93(1):48–53.

Lacombe, P., and D. Keyes. 2000. Defending the American Homeland's Infrastructure. *Journal of Homeland Defense* [Online]. Available: <http://www.homelanddefense.org>. [cited October 27, 2000]

Ladner, R.L., and G.A. Leake. 1999. Preparing Utility Computer Systems for the Year 2000. *Journal AWWA*, 91(2):38–45.

Lancaster-Brooks, R. 2002. Water Terrorism: An Overview of Water and Wastewater Security Problems and Solutions. In *Journal of Homeland Security* [Online]. Available: <http://www.homelanddefense.org/journal/articles/displayArticle>. [cited March 17, 2002]

Last, J.M. 1987. *Public Health and Human Ecology.* Stamford, Conn.: Appleton & Lange Canada.

Last, J.M. 1998. The History, Scope, and Methods of Public Health. In *Public Health and Human Ecology.* 2nd ed. Edited by J.M. Last. Stamford, Conn.: Appleton & Lange.

Latowsky, G.P. 1998. *Meeting Community Needs: Lessons Learned From Health Studies Conducted in Woburn, MA.* The Berkman Center at Harvard Law School [Online]. Available: <http://cyber.law.harvard.edu/acivilaction/MCN.htm>. [cited April 25, 2003]

LeChevallier, M.W. 1990. Coliform Regrowth in Drinking Water: A Review. *Journal AWWA*, 82:133–146.

LeChevallier, M.W., B.H. Olson, and G.S. McFeters. 1990. *Assessing and Controlling Bacterial Regrowth in Distribution Systems.* Denver, Colo.: AwwaRF.

Lennette, E.H., A. Balows, W.J. Hausler Jr., and H.J. Shadomy, eds. 1985. *Manual of Clinical Microbiology.* 4th ed. Washington, D.C.: American Society for Microbiology.

Letterman, R.D., ed. 1999. *Water Quality and Treatment.* 5th ed. New York: McGraw-Hill.

Levine, W.C., J.F. Smart, D.L. Archer, N.H. Bean, and R.V. Tauxe. 1991. Foodborne Disease Outbreaks in Nursing Homes, 1975 through 1987. *Journal of the American Medical Association*, 266:2105–2109.

Livernois, J. 2002. The Economic Costs of the Walkerton Water Crisis. In *The Walkerton Inquiry.* Commissioned Paper 14. Toronto, Ont.: Ontario Ministry of the Attorney General, Queen's Printer for Ontario.

Ljunggren, O. 1990. The User's Problem and Need for Corrosion Monitoring and Control. In *Corrosion and Corrosion Control in Drinking Water Systems.* Houston Texas: National Association of Corrosion Engineers.

Louie, J.N. 1999. *Geology 101 Course Notes.* Nevada Seismological Laboratory, University of Nevada, Reno [Online]. Available: <http://www.seismo.unr.edu/ftp/pub/louie/class/100/magnitude.html>. [cited April 1, 2002]

Luthy, R.G. 2001. Safety of Our Nation's Water. Statement to the Committee on Science. U.S. House of Representatives Hearing on H.R. 3178 and the Development of Anti-Terrorism Tools for Water Infrastructure, November 14, 2001 [Online]. Available: <http://www.aeesp.org/pubs/pfd/Safety_of_our_Water_Supply>. [cited April 1, 2002]

Macfarlane, M., and M. McIntosh. 2003. Stakeholder Consultation: Socializing Capital or Capitalising on the Social? [Online] Available: <http://users.wbs.warwick.ac.uk/ccu/research/bp_paper.pdf>. [cited April 25, 2003]

MacKenzie, W.R., N.J. Hoxie, M.E. Proctor, M.S. Gradus, K.A. Blair, D.E. Peterson, J.J. Kazmierczak, D.G. Addiss, K.R. Fox, J.B. Rose. 1994. A Massive Outbreak in Milwaukee of *Cryptosporidium* Infection Transmitted Through a Public Water Supply. *New England Journal of Medicine*, 331(22):1529–1530.

Mahoney, L.E., C.T.H. Friedman, R.A. Murray, E.L. Schulenburg, and G.A. Heidbreder. 1974. A Waterborne Gastroenteritis Epidemic in Pico Rivera, California. *American Journal of Public Health*, 64(10):963–968.

Marlow, H.A., and B.A. Beaudet. 1992. Workforce Literacy in the Water Utility Industry. *Journal AWWA*, 84(6):46–49.

Mays, L.W. 2000. Introduction. In *Water Distribution Systems Handbook*. Edited by L.W. Mays. New York: McGraw-Hill.

McElroy, A., and P.K. Townsend. 1989. *Medical Anthropology in Ecological Perspective*. 2nd ed. Boulder, Colo.: Westview Press.

McMichael, A.J. 2000. The Urban Environment and Health in a World of Increasing Globalization: Issues for Developing Countries. *Bulletin of the World Health Organization*, 78(9):1117–1126.

McReynolds, L., and R.L. Simmons. 1995. LA's Rehearsal for the Big One. *Journal AWWA*, 87(5):65–70.

Mims, C.A., J.H.L. Playfair, I.M. Roitt, D. Wakelin, and R. Williams. 1993. *Medical Microbiology*. London: Mosby Europe Ltd.

Moser, R.H. 2000. *Pathogen Intrusion Into Potable Water Distribution Systems* [Online]. Available: <http://www.asdwa.org/website/PowerPointPresentations/Pathogen%20Intrusion%20talk.doc>. [cited April 1, 2002]

Moyer, J. 2003. *The Roman Aqueducts and Water Systems* [Online]. Available: <http://academic.bowdoin.edu/classics/research/moyer/html/intro.shtm/>. [cited February 5, 2003]

Murphy, P.J. 1990. *Exposure to Wells G and H in Woburn Massachusetts* [Online]. Available: <http://www.state.ma.us/dph/beha/cau/reports/woburn>. [cited April 26, 2003]

National Drinking Water Clearinghouse. 1998. SDWA Requires Operator Certification. *On Tap: Drinking Water News for America's Small Communities*, 7:2.

NGA (National Governors Association). 2003. *Summary of the NGA Bioterrorism Summit* [Online]. Available: <http://www.nga.org/center/divisions>. [cited March 30, 2003]

NIH (National Institutes of Health). 1996. *Severe Weather Safety* [Online]. Available: <http://www.nih.gov/od/ors/newsltrs/psn/psn19.html>. [cited February 25, 2002]

NOAA (National Oceanic and Atmospheric Administration). 2002a. Beaufort Scale [Online]. Available: <http://www.crh.noaa.gov/lot/webpage/beaufort>. [cited March 10, 2002]

NOAA (National Oceanic and Atmospheric Administration). 2002b. Saffir-Simpson Scale [Online]. Available: <http://www.nhc.noaa.gov/HAW/basics/saffir_simpson.html>. [cited March 10, 2002]

Noji, E.K. 1991. The Role of Epidemiology in Natural Disaster Reduction: An Interdisciplinary Approach. In *Proc. of the Second US–Japan Natural Disaster Reduction Workshop*. Tokyo: Japan Science and Technology Agency.

Noji, E.K. 1997. The Use of Epidemiologic Methods in Disasters. In *The Public Health Consequences of Disasters*. Edited by E.K. Noji. New York: Oxford University Press.

NRC (National Research Council). 1983. *Risk Assessment in the Federal Government: Managing the Process*. Washington, D.C.: National Academy Press.

NTSB (National Transport Safety Board). 2003. *The Investigative Process* [Online]. Available: <http://www.ntsb.gov>. [cited April 1, 2003]

O'Conner, D.R. 2002. *Report of the Walkerton Inquiry: The Events of May 2000 and Related Issues*. Part 1. Toronto, Ont.: Ontario Ministry of the Attorney General.

Odgen, C.L., L.I. Gibbs-Scharf, M.A. Kohn, and J. Malilay. 2001. Emergency Health Surveillance After Severe Flooding in Louisiana, 1995. *Prehospital and Disaster Medicine*, 16(3):97–103.

OGC (Oil and Gas Commission). 2003. Consultation [Online]. Available: http://www.ogc.gov.bc.ca/consultation.asp>. [cited August 4, 2003]

Olson, B.H. 1982. *Assessment and Implications of Bacterial Regrowth in Water Distribution Systems*. EPA-600/52-82-072. Cincinnati, Ohio: USEPA.

Omenn, G.S., G.E. Goodman, M.D. Thornquist, J. Balmes, M.R. Cullen, A. Glass, J.P. Keogh, F.L. Meyskens Jr., B. Valanais, J.H. Williams Jr., S. Barnhart, M.G. Cherniack, C.A. Brodkin, and S. Hammar. 1996. Risk Factors for Lung Cancer and for Intervention Effects in CARET, the Beta-Carotene and Retinol Efficacy Trial. *Journal of the National Cancer Institute*, 88(21):1550–1559.

PAHO (Pan American Health Organization). 1998. *Natural Disaster Mitigation in Drinking Water and Sewerage Systems: Guidelines for Vulnerability Analysis*. Washington D.C.: PAHO.

PAHO (Pan American Health Organization). 2000a. *Natural Disasters: Protecting the Public's Health*. Washington, D.C.: PAHO.

PAHO (Pan American Health Organization). 2000b. *Principles of Disaster Mitigation in Health Facilities*. Washington, D.C.: PAHO.

Pearson, G., K. Martel, A. Hill, G. Burlingame, and A. Godfree. 2001. *Practices to Prevent Microbiological Contamination of Water Mains*. Denver, Colo.: AwwaRF and AWWA.

Peerenboom, J. 2002. *Infrastructure Interdependencies: Overview of Concepts and Terminology*. Argonne, Ill.: Argonne National Laboratory, Infrastructure Assurance Center.

Pew Environmental Health Commission. 2001. *Transition Report to the New Administration: Strengthening Our Public Health Defense Against Environmental Threats* [Online]. Johns Hopkins School of Public Health, Johns Hopkins University. Baltimore Md. Available: <http://healthyamericans.org/resources/files/transition.pdf>. [cited April 25, 2003]

Pierson, G., K. Martel, A. Hill, G. Burlingame, A. Godfree. 2001. *Practices to Prevent Microbiological Contamination of Water Mains*. Denver, Colo.: AwwaRF and AWWA.

PNWS–AWWA (Pacific Northwest Section–American Water Works Association). 1996. *Summary of Backflow Incidents*. 4th ed. Portland Ore.: PNWS–AWWA.

Pontius, F.W., and S.W. Clark. 1999. Drinking Water Quality Standards, Regulations, and Goals. In *Water Quality and Treatment*. 5th ed. Edited by R.D. Letterman. New York: McGraw-Hill.

Potter, M.E. 1999. Statement by Morris E. Potter, DVM, Director, Food Safety Initiatives, Center for Food Safety and Applied Nutrition, Food and Drug Administration, Department of Health and Human Services, Before the Subcommittee on Oversight of Government Management, Restructuring, and the District of Columbia Committee on Governmental Affairs, United States Senate [Online]. Available: <http://www.cfsan.fda.gov/~dms/steggs.html>. [cited May 19, 2002]

Poulsen, K. 2002. FBI Issues Water Supply Cyberterror Warning. In *SecurityFocus Home News* [Online]. Available: <http://online.securityfocus.com/news>. [cited April 1, 2002]

Prescott, S.C., and C.E.A. Winslow. 1904. *Elements of Water Bacteriology.* London: John Wiley & Sons.

Pruess A., D. Kay, L. Fewtrell, and J. Bartram. 2002. Estimating the Burden of Disease From Water, Sanitation, and Hygiene at a Global Level. *Environmental Health Perspectives*, 110(5):537–542.

Putnam, S.W., and J.D. Graham. 1993. Chemicals Versus Microbials in Drinking Water: A Decision Sciences Perspective. *Journal AWWA*, 85(3):57–61.

Reid, J. 1994. Overcoming the Flood: How Midwestern Utilities Managed Disaster. *Journal AWWA*, 86(1):58–67.

Reingold, A.L. 1998. Outbreak Investigation—A Perspective. *Emerging Infectious Diseases*, 14(1).

Riordan, R.A. 1995. Mutual Aid and Emergency Response for Water Utilities. *Journal AWWA*, 87(5):52–58.

Roberson, J.A., and C.T. Crowe. 1993. *Engineering Fluid Mechanics*. 5th ed. Boston, Mass.: Houghton Mifflin.

Ross, C.E., and J. Mirowsky. 2001. Neighborhood Disadvantage, Disorder, and Health. *Journal of Health and Social Behavior*, 42(9):258–276.

Roszak, D.B., and R.R. Colwell. 1987. Survival Strategies of Bacteria in the Natural Environment. *Microbiological Reviews*, 51:365–379.

Rowe, R.W. 1981. *A Retrospective Study of Woburn 1949–1981* [Online]. Available: <http://www.state.ma.us/dph/beha/cau/reports/woburn/retrowob/pdf>. [cited April 25, 2003]

Ruggenini, A.M., M.C. Pastoris, P.J. Dennis, C. Barral, A. Sciacovelli, F. Carle, M. Bolgiani, G. Passarion, M.G. Mingrone, C. Passi, and M. Lombardo. 1989. *Legionella pneumophila* in a Hospital in Torino, Italy. A Retrospective One-Year Study. *Epidemiology and Infection*, 102:21–29.

Saxe, J.G. [N.d.] The Blind Men and the Elephant [Online]. Available: <http://www.wvu.edu/~lawfac/jelkins/1p-2001/saxe.html>. [cited January 2, 2003]

Scala, R.A. 1991. Risk Assessment. In *Casarett and Doull's Toxicology.* 4th ed. Edited by M.O. Amdur, J. Doull, and C.D. Klaassen. New York: Pergamon Press.

SCEDC (Southern California Earthquake Data Center). 2002. *San Fernando Earthquake* [Online]. Available: <http://www.scecdc.scec.org/sanfer.html>. [cited March 1, 2002]

Schock, M.R. 1999. Internal Corrosion and Deposition Control. In *Water Quality and Treatment.* 5th ed. Edited by R.D. Letterman. New York: McGraw-Hill.

Schoenen, D., and B.D. Schoeler. 1985. *Drinking Water Materials: Field Observations and Methods of Investigation*. Chichester: Ellis Horwood.

Science and Environmental Health Network, Johnson Foundation, W. Alton Jones Foundation, C.S. Fund, and Lowell Center for Sustainable Production. 1998. Wingspread Conference on the Precautionary Principle, January 26, 1998 [Online]. Available: <http://www.sehn.org/web2printer4.php>. [cited May 20, 2002]

Semenza, J.C., L. Roberts, A. Henderson, J. Bogan, and C.H. Rubin. 1998. Water Distribution System and Diarrheal Disease Transmission: A Case Study in Uzbekistan. *American Journal of Tropical Medicine and Hygiene*, 59(6):941–946.

Sidel, V.W., E. Onel, H.J. Geiger, J. Leaning, and W.H. Foege. 1992. Public Health Responses to Natural and Human-Made Disasters. In *Public Health and Preventive Medicine*. 13th ed. Edited by J.M. Last and R.B. Wallace. Toronto, Ont.: Appleton & Lange.

Snyder, J.K., A.K. Deb, F.M. Grablutz, S.B. McKinnon, W. Grayman, R.M. Clark, D.A. Okun, S.M. Typer, and D. Savic. 2002. *Impacts of Fire Flow on Distribution System Water Quality, Design, and Operation*. Denver, Colo.: AwwaRF and AWWA.

Somerville, P. 1995. Kobe Earthquake: An Urban Disaster. *Eos*, 76(6):49–51.

SPC (Storm Prediction Center). 2002. Fujita Tornado Scale [Online]. Available: <http://www.spc.noaa.gov/archive/tornados>. [cited March 10, 2002]

St. Petersburg Times. 2001. 2 Water Main Breaks Cut Service to 350 Homes [Online]. Available: <http://www.sptimes.com/News/072101/Pasco/2_water-main_breaks>. [cited April 8, 2002]

Stevenson, L.H. 1978. A Case for Bacterial Dormancy in Aquatic Systems. *Microbial Ecology*, 4:127–122.

Stokes III, J., J.J. Noren, and S. Shindell. 1982. Definition of Terms and Concepts Applicable to Clinical Preventative Medicine. *Journal of Community Health*, 8:33–41.

Sutherland, I.W. 1989. Microbial Exopolysaccharides—Their Role in Microbial Adhesion in Aqueous Systems. *CRC Critical Reviews in Microbiology*, 10:173–201.

Sylvester, J.M., T.M. Reiter, and J.T. Waldron III. 2000. Insurance Issues Surrounding Pipe-Related Liabilities. *Journal AWWA*, 92(7):40–49.

Talbot, J.A., A. Barry, and N. Fok. 2001. Setting Up a Community-Based Surveillance Program—A Watchdog Approach. In *Waterborne Gastrointestinal Disease Outbreak Detection*. Edited by K.M.E. Emde, J.A. Talbot, L. Gammie, J. Mainiero, E. Geldreich, A. Barry, N. Fok, and B. Reilley-Matthews. Denver, Colo.: AwwaRF and AWWA.

Tanaka, S.S. 1995. Shaken Into Action. *Journal AWWA*, 87(5):71–80.

Tarr, J.A. 1987. Water Supply History. In *History of Pollution in Woburn, Massachusetts* [Online]. Available: <http://www.civil-action.com/facts/keydocuments/tarrrpt.html>. [cited April 26, 2003]

Thacker, S.M., and R.L. Berkelman. 1988. Public Health Surveillance in the United States. *Epidemiologic Reviews*, 10:164–190.

Thomson, M., and EnviroChex Associates. 2002. Exercising Due Diligence—In Municipal Water Systems [Online]. Available: <http://oakvillenewsonline.com/LEGAL.FCM_Feb902_DueDiligence_FCM2.pdf>. [cited April 18, 2003]

3M Company. 2002. History [Online]. Available: <http://www.3m.com>. [cited April 15, 2002]

Tillett, H.E., J. de Louvois, and P.G. Wall. 1998. Surveillance of Outbreaks of Waterborne Infectious Disease: Categorizing Levels of Evidence. *Epidemiology and Infection*, 120:37–42.

Tilling, R.I., L. Topinka, and D.A. Swanson. 1990. *Eruptions of Mount St. Helens: Past, Present, and Future*: USGS Special Interest Publication [Online]. Available: <http://vulcan.wr.usgs.gov/Volcanoes/MSH>. [cited March 1, 2002]

USEPA (United States Environmental Protection Agency). 1997. *Drinking Water Infrastructure Needs Survey: First Report to Congress.* EPA 812-R-97-001. Washington, D.C.: USEPA.

USEPA (United States Environmental Protection Agency). 2003. *Draft Final Guidelines for Carcinogen Risk Assessment* [Online]. Available: <http://cfpub.epa.gov/ncea/recordisplay.cfm>. [cited May 22, 2002]

USFDA (United States Food and Drug Administration). 2001. *Bacteriological Analytical Manual* [Online]. Available: <http://www.cfsan.fda.gov/~ebam/bam-mm.html>. [cited May 29, 2002]

USGS (United States Geological Survey). 2002. Chile May 22 Earthquake [Online]. Available: <http://neic.usgs.gov/neis/eqlists/WORLD/1960_05.22.html>. [cited March 1, 2002]

Valanis, B. 1992. *Epidemiology in Nursing and Health Care.* 2nd ed. Norwalk, Conn.: Appleton & Lange.

van der Kooij, D. 1988. Prevention of Bacterial Aftergrowth in Drinking Water Distribution Systems. *Water Supply,* 6:13–18.

van der Kooij, D. 1990. Assimilable Organic Carbon (AOC) in Drinking Water. In *Drinking Water Microbiology.* Edited by G.A. McFeters. New York: Springer-Verlag.

Vik, E.A., R.A. Ryder, I. Wagner, and J.F. Ferguson. 1996. Mitigation of Corrosion Effects. In *Internal Corrosion of Water Distribution Systems.* Denver, Colo.: AwwaRF and DVGW-Technologiezentrum Wasser.

Walski, T.M., and T.L. Lutes. 1994. Hydraulic Transients Cause Low-Pressure Problems. *Journal AWWA,* 86(12):24–32.

Warburton, D.W. 1993. A Review of the Microbiological Quality of Bottled Water Sold in Canada. Part 2. The Need for More Stringent Regulations. *Canadian Journal of Microbiology,* 39:158–168.

Wasley, A. 1995. Epidemiology in the Disaster Setting. *Current Issues in Public Health,* 1:131–135.

Wetterhall, S.F., and E.K. Noji. 1997. Surveillance and Epidemiology. In *The Public Health Consequences of Disasters.* Edited by E.K. Noji. New York: Oxford University Press.

Whelan, E.M. 1993. *Toxic Terror: The Truth Behind the Cancer Scares.* Buffalo, N.Y.: Prometheus Books.

WHO (World Health Organization). 1996. *Guidelines for Drinking Water Quality.* 2nd ed. Vol. 2: Health Criteria and Other Supporting Information. Geneva: WHO.

WHO (World Health Organization). 2003. WHO Definition of Health [Online]. Available: <http://www.who.int/about/definition/en>. [cited April 18, 2003]

Williams, B.L., and Y. Florez. 2002. Do Mexican Americans Perceive Environmental Issues Differently Than Caucasians: A Study of Cross-Ethnic Variation in Perceptions Related to Water in Tucson. *Environmental Health Perspectives,* 110 (suppl. 20):303–310.

Wilson, I.B., and P.D. Cleary. 1995. Linking Clinical Variables With Health-Related Quality of Life: A Conceptual Model of Patient Outcomes, *Journal of the American Medical Association,* 273(1):59–65.

Wiltshire, C.E. 1997. Managing Risk and Risk Acceptance. *Optimum, the Journal for Public Sector Management,* 27(3):14–23.

Wise, D. 1998. Cross-Training Benefits Oregon Plant. *Journal AWWA,* 90(10):60–66.

Xie, Y., C.A. Cole, and D.A. Long. 1999. Training Center Targets Small System Operators. *Journal AWWA,* 91(4):123–128.

Youd, T.L. 1993. Liquefaction, Ground Failure and Consequent Damage During the 22 April 1991 Costa Rica Earthquake [Online]. Available: <http://nicee.berkeley.edu/costarica>. [cited May 20, 2002]

Zaillian, S. (director). 1999. *A Civil Action*. Burbank, Calif.: Touchstone Pictures.

ABBREVIATIONS

AOAC	AOAC International
APHA	American Public Health Association
ASME	American Society of Materials Engineering
ASQ	American Society for Quality
ASTDHPPHE	Association of State and Territorial Directories of Health Promotion and Public Health Education
ATSDR	United States Agency for Toxic Substances and Disease Registry
AWWA	American Water Works Association
AwwaRF	Awwa Research Foundation
BEHA	Bureau of Environmental Health Assessment
°C	degrees Celsius
CDC	Centers for Disease Control and Prevention
cfu/mL	colony forming units per milliliter
CIPHI	Canadian Institute of Public Health Inspectors
cm	centimeter
DDT	dichloro-diphenyl-trichloroethane
DNA	deoxyribonucleic acid
DNR	Department of Natural Resources (Massachusetts)
DoD	Department of Defense (United States)
DOH	Department of Health (Massachusetts)
DOT	Department of Transportation (United States)
DS	distribution system
ECT	emergency communications team
EMS	Emergency Medical Services
°F	degrees Fahrenheit
FAA	Federal Aviation Agency (United States)
FBI	Federal Bureau of Investigation (United States)
FDA	Food and Drug Administration (United States)
FEMA	Federal Emergency Management Agency (United States)
FSIS	Food Safety and Inspection Service
ft	foot
gal	gallon
GASB 34	Governmental Accounting Standards Board, Statement 34
GIS	geographic information system
gpm	gallons per minute
GVRD	Greater Vancouver (Canada) Regional District

HPC	heterotrophic plate count
hr	hour
IAMFES	International Association of Milk, Food and Environmental Sanitarians
in.	inch
ISO	International Organization for Standardization
JIC	joint information center
kPa	kilopascal
L	liter
LIMS	Laboratory Information Management System
L/sec	liters per second
m	meter
M.D.	medical doctor
MIC	microbially influenced corrosion
mL	milliliter
mm	millimeter
mph	miles per hour
μg/L	micrograms per liter
NATO	North Atlantic Treaty Organisation
NGA	National Governors Association (United States)
NIH	National Institutes of Health (United States)
NOAA	National Oceanic and Atmospheric Administration (United States)
NRC	National Research Council
NSF	NSF International
NTSB	National Transport Safety Board (United States)
OGC	Oil and Gas Commission
O&M	operation and maintenance
PAHO	Pan American Health Organization
PE	polyethylene
P.E.	professional engineer (United States)
P. Eng.	professional engineer (Canada)
PHD	public health department
PNWS–AWWA	Pacific Northwest section of the American Water Works Association
ppb	parts per billion
ppm	parts per million
psi	pounds (force) per square inch
PVC	polyvinyl chloride

SARS	severe acute respiratory syndrome
SCADA	supervisory control and data acquisition
SCEDC	Southern California Earthquake Data Center
SOP	standard operating procedure
SPC	Storm Prediction Center
THM	trihalomethane
T&O	taste and odor
TQM	Total Quality Management
24/7	24 hours per day, 7 days per week
U.S.	United States
USEPA	United States Environmental Protection Agency
USGS	United States Geological Survey
UV	ultraviolet
UV_{254}	ultraviolet 254 nanometers
WEF	Water Environment Federation
WHO	World Health Organization
WTP	water treatment plant
Y2K	year 2000 (millennium)

IWA Publishing's authorised EU representative for General Product Safety Regulations is Diane D'Arras, 15 rue Duret, 75116 Paris, France, e-mail: safety@iwap.co.uk.

Printed and bound by CPI Group (UK) Ltd, Croydon, CR0 4YY
12/05/2026
02108865-0017